T0074836

# Rapid Review of Chemistry for the Life Sciences and Engineering

# Rapid Review of Chemistry for the Life Sciences and Engineering

## With Select Applications

Armen S. Casparian, Gergely Sirokman,
Ann O. Omollo

CRC Press
Taylor & Francis Group
Boca Raton London New York

CRC Press is an imprint of the
Taylor & Francis Group, an **informa** business

First edition published 2022
by CRC Press
6000 Broken Sound Parkway NW, Suite 300, Boca Raton, FL 33487-2742

and by CRC Press
2 Park Square, Milton Park, Abingdon, Oxon, OX14 4RN

© 2022 Armen S. Casparian, Gergely Sirokman, Ann Omollo

CRC Press is an imprint of Taylor & Francis Group, LLC

Reasonable efforts have been made to publish reliable data and information, but the author and publisher cannot assume responsibility for the validity of all materials or the consequences of their use. The authors and publishers have attempted to trace the copyright holders of all material reproduced in this publication and apologize to copyright holders if permission to publish in this form has not been obtained. If any copyright material has not been acknowledged please write and let us know so we may rectify in any future reprint.

Except as permitted under U.S. Copyright Law, no part of this book may be reprinted, reproduced, transmitted, or utilized in any form by any electronic, mechanical, or other means, now known or hereafter invented, including photocopying, microfilming, and recording, or in any information storage or retrieval system, without written permission from the publishers.

For permission to photocopy or use material electronically from this work, access www.copyright. com or contact the Copyright Clearance Center, Inc. (CCC), 222 Rosewood Drive, Danvers, MA 01923, 978-750-8400. For works that are not available on CCC please contact mpkbookspermissions@ tandf.co.uk

*Trademark notice*: Product or corporate names may be trademarks or registered trademarks and are used only for identification and explanation without intent to infringe.

*Library of Congress Cataloging-in-Publication Data*
Names: Casparian, Armen S., author.
Title: Rapid review of chemistry for the life sciences and engineering :
  with select applications / Armen S. Casparian, Gergely Sirokman, Ann
  Omollo.
Description: Boca Raton : CRC Press, [2021] | Includes bibliographical
  references and index.
Identifiers: LCCN 2021023350 (print) | LCCN 2021023351 (ebook) | ISBN
  9780367541668 (hardback) | ISBN 9780367552794 (paperback) | ISBN
  9781003092759 (ebook)
Subjects: LCSH: Chemistry.
Classification: LCC QD31.3 .C357 2021  (print) | LCC QD31.3 (ebook) | DDC
  540—dc23
LC record available at https://lccn.loc.gov/2021023350
LC ebook record available at https://lccn.loc.gov/2021023351

ISBN: 978-0-367-54166-8 (hbk)
ISBN: 978-0-367-55279-4 (pbk)
ISBN: 978-1-003-09275-9 (ebk)

DOI: 10.1201/9781003092759

Typeset in Times
by codeMantra

*ASC wishes to dedicate this book to his children: Brendan, Dara, and Meghan. I hope they read this book (or at least parts of it) and satisfy some of their intellectual curiosity. I hope they go on to inspire others about consumer awareness and choices, as well as environmental quality, be it in the workplace, marketplace, at home, or in lands yet to be explored.*

*AOO dedicates this book to her late father, Jaduong' Ondera: "So grateful that you gave me equal opportunity to education at a time when educating girls was considered debatable, Dad."*

### *THANKS*

*ASC also wants to take this opportunity to thank Professor Jerry Hopcroft, colleague, editor, and friend, and Joel Stein, editor and friend, whose unwavering support, insights, and leadership over many years helped inspire and complete this book. There have also been other colleagues and close friends (MG, SK, IG, and MK) who have inspired me in different ways.*

*To All Our Reviewers and Editors, Hilary and Jessica, Iris and Saranya—Many, Many Thanks!*

# Contents

13.7    Environmental Fate and Transport of Selected
        Water Pollutants................................................................ 184
        13.7.1   DDT.................................................................... 184
        13.7.2   Pharmaceutical Pollution ...................................... 185
        13.7.3   Dioxins ............................................................... 185

**Chapter 14**   The Chemistry of Hazardous Materials......................................... 187

        14.1    Background........................................................................ 187
        14.2    The Chemistry of Four Common Elements ......................... 189
                14.2.1   Hydrogen .............................................................. 189
                14.2.2   Oxygen ................................................................. 191
                14.2.3   Chlorine................................................................ 193
                14.2.4   Sulfur.................................................................... 195
        14.3    The Chemistry of Some Corrosive Materials..................... 196
                14.3.1   The Nature of Corrosivity ..................................... 196
                14.3.2   The Nature and Properties of Acids and Bases........ 197
                14.3.3   Select Weak Acids of Interest ................................ 198
                14.3.4   Select Bases of Interest.........................................200
                14.3.5   Anhydrides of Acids and Bases .............................200
        14.4    The Chemistry of Pyrophoric Substances...........................201
        14.5    The Chemistry of Flammable Substances...........................203
                14.5.1   OSHA Classification System..................................203
                14.5.2   NFPA Classification System .................................204
        14.6    The Chemistry of Explosives ...........................................205
        14.7    Green Chemistry ...............................................................208
        14.8    Nanotechnology................................................................ 210
        14.9    Three Important "Environmental" Disasters ....................... 212
                14.9.1   West Virginia MCHM Leak of 2014...................... 212
                14.9.2   British Petroleum and the *Deepwater Horizon*
                         Accident of 2010................................................... 214
                14.9.3   Union Carbide and Bhopal, India of
                         1984—The Worst Chemical Accident Ever ............. 215

**Chapter 15**   Introduction to Basic Toxicology ................................................. 219

        15.1    The Problem ..................................................................... 219
                15.1.1   Five Immediate Conclusions .................................222
        15.2    Basic Concepts .................................................................223
        15.3    More Terms and Definitions...............................................227
        15.4    The Dose–Response Relationship .......................................230
        15.5    Multiple Chemical Sensitivity (MCS) Syndrome................233
        15.6    Pesticides: Two Examples—Atrazine, Arsenic....................234
                15.6.1   Atrazine ................................................................235
                15.6.2   The Persistence of Arsenic—A Crossover
                         Chemical—A Pesticide and More...........................236

# Foreword

Why another book on chemistry? In a nutshell, it is first and foremost to provide a quick and easy reference manual written in clear and plain language for practitioners of chemistry who need to review and refresh their understanding of the fundamental principles. Secondarily, armed with select, contemporary applications, it is written in the hope to bridge the gap between chemists and non-chemists, so that they may communicate with and understand each other. It's fair to assume that many pressing technological problems in the immediate future will require solutions formulated by teams of experts composed of not only scientists (including chemists), but also engineers, medical doctors, health and safety professionals, economists, financial managers, and government officials. The non-chemists or nonscientists on the team will have to be able to communicate with the chemists and understand their language to arrive at mutually acceptable and optimal solutions to complex problems of enormous consequences.

Education, whether formal or self-taught, often leads to a thirst for more knowledge. It promotes the ability to see patterns of information from diverse areas and integrate them more and more quickly. Native intelligence is the spark that lights the fire for education. After all, whether we admit it or not, all of us are involved in a search for how and why did we get here.

In 2019, 3M's State of Science Index (C&EN; April 6, 2020) surveyed more than 14,000 people in 14 countries exploring the themes of the appreciation and trust of science, the image of scientists, and the importance of science in policy setting and decision-making in government. Here are some of what they found. When asked to rate how important science is to society today, 64% in the United States said it was very important while 82% in Brazil said it was. When asked the question "When you think about the role of science in the next 20 years, does it make you feel more trusting or suspicious?," 85% of South Koreans answered "trusting" compared with only 67% of US respondents. The response rate may be influenced by the fact that South Korea was the highest spending country in the world in terms of R&D as a percentage of gross domestic product—4.3% of GDP. The US's R&D spending is 2.7% by comparison, while China's is only 2%, despite the fact that 88% of Chinese respondents described themselves as trusting of the role of science in the future.

It is evident that life is both fragile and yet tenacious. The recent pandemic of COVID-19 caused by SARS CoV-2 is one example of this. So are the tiny weeds sprouting in between brick pavers in a backyard patio or walkway, which can be hand-picked or sprayed with an herbicide, but nevertheless reappear a short time later. We would like the reader to keep that struggle in mind when reading the latter chapters of this book, with applications including exposure to hazardous chemicals and their toxicology in their relation to consumer products and human health. For the impatient and intellectually curious reader, we suggest watching a short, 7-minute video clip on YouTube entitled "Little Things Matter" narrated by Dr. Bruce Lamphear and produced by the Canadian Broadcasting Company (CBC) in 2016. It is outstanding!

Of more immediate consequence, however, the American economy, perhaps the world economy, centers on technological advancements hinged on science and engineering. Chemistry is the central science and the bridge that connects the natural sciences to one another and to engineering, and ultimately to technological advances. To say that everyone in today's world should understand some level of chemistry, be that person an engineer, life scientist, businessman, or informed consumer, is an understatement.

There is considerable concern and worry for the survival of our planet, which necessarily means the survival of our species, the human race. Survival rests on sustainability. Sustainability depends on a balance of the three factors—world economy, global society cooperation, and the physical environment. Sustainability opposes growth for the sake of growth—the philosophy of the cancer cell. Put another way, not all change is progress. At this time, the world economy appears to be on a collision course with the integrity of the physical environment.

Many observers are concerned that apex predators and diversity of species are disappearing; coral reefs are dying in acidified oceans; glacial water supplies are melting and raising sea levels; and land is being deforested and desertified at incredible rates. The use of synthetic chemicals in manufacturing and commerce is increasing every year. Human ingenuity has solved many problems. Consider, for example, how efficient man has become at catching fish from the ocean, to the extent that it is depleting breeding stocks, leading to collapse of fish populations. This has required international legislation to regulate who can fish where and when, and how much can be legally caught. Does anyone know if these laws are being enforced? Who are the enforcing agencies? William Forster Lloyd, a 19th-century mathematician who first developed the "Tragedy of the Commons," could be turning over in his grave.

Our civilization—location of our cities, the crops we grow, the technologies that underpin our industry and commerce—is based on a climate pattern that is rapidly changing. Scientific evidence is predicting that the Earth will become a warmer planet, sea levels will rise, precipitation patterns will shift causing many areas to suffer from drought, while other areas to experience floods. Temperature extremes will become more common. Biodiversity will probably decrease, and new kinds of pests and pathogens may develop and flourish, requiring new types of pesticides.

The physical environment means air and water quality and pollution, climate and its changes, energy production and consumption, and ultimately, all of the Earth's resources. How they are interconnected and managed will determine changes in our lifestyles. Who will make these decisions?

To understand, maintain, and protect the physical environment, a basic understanding of chemistry, biology, and physics, and their hybrids is useful. Civil and environmental engineers, perhaps more than any other single engineering group, will be relied upon to solve our problems, design solutions, and help escort us into a new "normal." But the wisdom and expertise of nonscientifically educated individuals will also be needed and equally treasured. This, then, is the motivation behind this book. We hope that individuals with interest from all disciplines and backgrounds will find this book readable and useful. As Mark Twain once observed: "History may not always repeat itself, but it sure does rhyme a lot."

# Preface

This book is intended to serve as a reference manual that demystifies chemistry for the non-chemist who, nevertheless, may be a practitioner of some area of science or engineering requiring or involving chemistry. It provides quick and easy access to fundamental chemical principles, quantitative relationships, and formulas.

Chapters 1–10 are designed to contain the standard material in an introductory college chemistry course. They are designed to refresh the reader's memory and provide a basic review of the fundamental principles of chemistry. It is assumed that the reader has already taken a one- or two-semester course in introductory chemistry at the college level, with a brief exposure to some organic and polymer chemistry. If not, the motivated reader with

Liquid nitrogen flowing into a Dewar flask: Transfer temperature between 77 and 63 K (–326 to –340 degrees F).

intellectual curiosity can still learn the basics of these two specialized topics from Chapters 8 and 9 for organic chemistry, and from Chapter 10 for polymer chemistry. The authors have made every effort to use language, consistency in symbols and notation, well-drawn and labeled figures, and up-to-date data tables, along with explanations of complex topics that are student-friendly. More than 100 solved examples are presented. For material on quantum chemistry and its applications, the interested reader may wish to consult a college-level chemistry textbook.

Chapters 11–15 present "applications of chemistry" that should interest and appeal to scientists and engineers engaged in a variety of fields. More specifically, the chapter subjects discussed are as follows:

- Chapter 11: Radioactivity and Nuclear Chemistry
- Chapter 12: The Atmosphere and the Chemistry of the Air and Air Pollution; Indoor Air Quality Factors
- Chapter 13: Water Quality and Water Pollution
- Chapter 14: The Chemistry of Hazardous Materials
- Chapter 15: An Introduction to Basic Toxicology, Including Some Discussion of Consumer Products and Their Risk to Human Health

# Authors

**Armen S. Casparian** completed his B.A. in Chemistry at Rutgers University and his M.Sc. (ABD) in Physical Chemistry at Brown University. He has been a Professor of Chemistry at Wentworth Institute of Technology for 33 years (Retired June 2013), where he developed and taught chemistry courses to non-chemistry majors. Since 2014, he has been an Adjunct Professor of Chemistry at the Community College of Rhode Island. He is also a Chemical and Engineering Consultant in private practice with over 40 years of experience and expertise in: designing and teaching applied and engineering chemistry courses; laboratory instruction in analytical methods; chemical health and safety issues; compliance with OSHA and EPA regulations; product liability issues involving paints, coatings and adhesive failure; and surfactant chemistry. He is a Registered Professional Chemist (ACS, 1968), Certified Chemical Hygiene Officer (NRCC, 1998), and Registered Professional Industrial Hygienist (APIH, 1995). He has published more than 75 review articles of textbooks, reference manuals, and multimedia presentations on topics including chemical health and safety, environmental quality, climate change, and management of hazardous materials.

**Gergely Sirokman** is a Professor of Chemistry in the School of Sciences and Humanities at Wentworth Institute of Technology where he has taught since 2008. He earned a B.S. in chemistry from Brandeis University and a PhD in inorganic chemistry from MIT. He has extensive experience in teaching chemistry to non-science majors, especially engineers and construction management students. He has personally mentored many students working on a variety of projects related to his academic interests. He has focused on the gamification of education and has designed games to teach science concepts at the college level. He also has a deep interest in renewable energy and climate change.

**Ann O. Omollo** completed her Bachelor of Education (Science) at Egerton University and PhD in Chemistry at Miami University. She is a Professor of Chemistry at Community College of Rhode Island (2009–present). She has taught courses in Organic Chemistry I & II, General Chemistry I & II, Chemistry of Our Environment, and Basic Skills. She has worked with many Honors students at CCRI over the years. She is also a Registered Professional Chemist (ACS, since 2003).

# For Quick Reference/Access to Solved Examples by Chapter Topic and Number

# 1 Introduction

## 1.1 THE BASICS OF MATTER

Chemistry is the study of the behavior of matter, particularly as different kinds of matter interact with each other on the very small scale of atoms. Matter is composed of atoms, which are generally referred to as indivisible particles and imagined as tiny billiard balls. Atoms are very small, being objects between 0.1 and 0.5 nm in diameter.

Atoms are, despite their assumed "indivisibility," composed of even smaller particles. (Atoms are in fact divisible, and the consequences of this fact are discussed in Chapter 11 of this book.) The particles that compose an atom are protons and neutrons, collectively known as nucleons, and electrons. Protons are positively charged particles and neutrons are electrically neutral particles and together carry the vast majority of the mass of the atom, whereas electrons are negatively charged particles and occupy most of the volume. Protons and neutrons have approximately the same mass while electrons are approximately 1/2000 the mass of either the proton or neutron.

Individual atoms are, in general, highly reactive and unstable, and so they tend to gather into groups known as molecules which lead to the formation of compounds. Compounds have chemical behaviors and physical properties that are distinct from the behavior of their component atoms. For example, ordinary table salt or sodium chloride has nothing in common with either sodium or chlorine, each of which is composed of a highly hazardous, reactive atom. Compounds are generally composed of simple whole number ratios of atoms combined into a new unit. The ways atoms form compounds can be broadly split into two categories, those being through covalent bonding, to make molecules where bonding electrons are shared, or ionic bonding, to make ionic compounds where electrons are effectively transferred from one atom to another. Other types of bonding and materials do exist, such as metallic bonding, necessary for the conduction of electrons in pure metals) and covalent network materials, (e.g. graphene and/or semi-conductors) but most of the concerns here will be covered by molecular and ionic compounds.

Perhaps, more to the point, chemistry is based on the premise that all matter is composed of some *combination* of 92 naturally occurring elements. Of these 92 elements, 11 are gases and 2 are known liquids at room temperature and pressure. Two others are solids but turn liquid at body temperature. The vast majority are solids and metals. The word "combination" is the key concept that gives chemistry its reputation of complexity. It may mean a mixture of elements, a mixture of compounds, or mixtures of elements with compounds. Furthermore, mixtures may be subdivided into homogeneous or heterogeneous. And to make matters even more complex, these systems of mixtures, if solutions, may exist as solids, liquids, or gases. Part of the challenge in chemistry, then, is to sort things out, organize, characterize, and identify them to understand their physical and chemical properties.

At the risk of repetition, chemistry is the branch of physical science that studies matter and the changes or "molecular rearrangements" it can undergo. These rearrangements,

DOI: 10.1201/9781003092759-1

conventionally known as chemical reactions, involve the breaking of existing chemical bonds between atoms to form new bonds with other atoms. In the process, different molecules with different properties are formed. Electrons may be gained or lost, and energy changes generally accompany the reaction. Chemical reactions, as distinguished from nuclear reactions, involve the exchange of only electrons— never protons or neutrons. In a cosmic sense, chemistry is the recycling of atoms. Visible evidence of chemical reactions includes the following:

- Bubbling or fizzing, indicating the release of a gas
- Color change
- Temperature change (heat released or absorbed)
- Formation of a precipitate
- Emission of light (chemiluminescence) or sound

Examples of common chemical reactions include the following:

- Rusting of iron or corrosion of any metal
- Generation of a current by a battery
- Combustion of fuel to produce energy
- Neutralization of excess stomach acid by an antacid
- Hardening of concrete

Chemistry also studies the structure of matter, including chemical and physical properties, correlating properties on the microscopic scale with behavior observed on the macroscopic scale. Included are such properties as vapor pressure, osmotic pressure, solubility, boiling and melting points, and energy and its transformations. Many of these properties may dictate or influence the outcome of a reaction.

There are actually four, conventional, physical states in chemistry: solid (s), liquid (l), gas (g), and aqueous solution (aq). The first three are regarded as pure, consisting of only a single substance. For example, $CaCl_2(s)$ would be pure, solid calcium chloride, while $CaCl_2$ (l) would be molten calcium chloride (at a very high temperature). The fourth physical state, *solution*, is a homogeneous mixture, i.e., of uniform and constant composition, where a solute and a solvent can be identified and distinguished. While attention to notation are may appear academic, it is essential and useful in understanding what products were actually formed in a chemical reaction. There are gaseous solutions like air (oxygen dissolved in nitrogen), liquid solutions like salt water (sodium chloride dissolved in water), and solid solutions like gemstones (iron or chromium atoms regularly spaced in an aluminum oxide crystal to form a sapphire or ruby, respectively). Diamond is composed of pure carbon atoms and is therefore not a solid solution. It is an allotrope of carbon; fullerene and graphite are also allotropes of carbon.

Often referred to as the universal solvent, water is a ubiquitous substance on the Earth and in the human body, and is the basis of all known life. Water-based or aqueous solutions are given the symbol (aq) inserted immediately after the chemical formula. Thus, an aqueous solution of calcium chloride would be written as $CaCl_2$ (aq), indicating that a given amount of solid calcium chloride was dissolved in a sufficient volume of water to make an aqueous solution. Note that water alone is a solvent, not a solution. The presence of the symbol (aq) always indicates the presence of a solute (usually a solid, but may also be another liquid or gas) dissolved in water, the solvent, to form a solution.

## 1.2 THE PERIODIC TABLE: BASIC CONCEPTS, SYMBOLS AND NOTATION, AND COMMON QUANTITIES AND THEIR UNITS

### 1.2.1 SYMBOLS AND NOTATION

A. As previously mentioned, chemistry is based on the premise that all matter is composed of some combination of the 92 naturally occurring elements. [Although the periodic table lists 118 elements, the final 26 are man-made or artificially synthesized, radioactive, and increasingly unstable, i.e., they have short half-lives. They are not addressed in this chapter, but are discussed in Chapter 11.] These elements are arranged in a particular order, known as **the Periodic Chart** or **Periodic Table**, in horizontal rows called **periods**, by increasing atomic number, and in vertical columns called **groups or families**, by similar electron arrangements or configurations, which give rise to similar chemical properties. **Electron configurations** are ordered arrangements of electrons based on specific housing rules. Energy levels or shells are quantized and exist from one to infinity. In the ground state of an atom, however, electrons can populate energy levels ranging from 1 to 7. Within each energy level or shell are subshells containing orbitals, the first four of which are labeled $s$, $p$, $d$, and $f$. Each of the four orbitals has a distinct shape and population limit: $1 s$ orbital with a maximum of 2 electrons; $3 p$ orbitals with a maximum of 6 electrons; $5 d$ orbitals with a maximum of 10 electrons; and $7 f$ orbitals with a maximum of 14 electrons.

B. Members of the same family are called **congeners**, especially important in organizing organic compounds such as polychlorinated biphenyls (or PCBs) or organochlorine pesticides. Here, the PCB molecule may have two, four, six, or eight chlorine atoms; all are congeners of the same family. Of more immediate relevance is the fact that some rows of the Periodic Table have names, such as the actinide (row 6) or lanthanide (row 7) series, while columns have names such as alkali metals (column 1A or 1), alkaline-earth metals (column 2A or 2), transition metals (columns 3 through 12), halogens (column 7A or 17), and noble (inert) gases (column 8A or 18). The pure numeral system of identifying columns is more modern and preferred over the number and letter system.

C. Each element is symbolized by either a single capital letter or a capital letter followed by a single lower-case letter. Thus, the number of elements from which a compound is made up can easily be determined by counting the number of capital letters.

D. The **atomic number**, **Z**, of an element is the number of protons in the nucleus of its atom. The atomic number, which is always an integer and uniquely identifies the element, is written as a *left-hand subscript* to the element symbol, for example, $_6$C.

E. The **mass number**, **A**, of an element is the sum of the number of protons and the number of neutrons in the nucleus, collectively known as the number of nucleons. Protons and neutrons are assigned a mass number of one atomic mass unit (amu) each, based on the carbon-12 atom as the standard. Although an element can have only one atomic number, it may have more than one mass number. This fact gives rise to the phenomenon of isotopes. **Isotopes** are atoms of the same element with the same atomic number but with a different number of neutrons. An element may have several isotopes, one or more of which may be radioactive and hence unstable. For example, oxygen has three naturally occurring isotopes, all of which are stable, while carbon also has three, one of which is radioactive. The mathematical average of all isotopic mass numbers of an element, weighted by the percent abundance in nature of these isotopes, constitutes the element's **atomic mass**. It is impossible to predict theoretically how many isotopes an element may have or how many may be radioactive. The isotopes of a given element have almost identical chemical properties (i.e., reactivity) but different physical properties (i.e., density, melting point, etc.). The mass number is expressed as a *left-hand superscript* to the element symbol, for example, $^{12}C$.

F. In its elemental state, the electronic charge of an element is zero; that is, the number of electrons equals the number of protons. If an atom gains or loses electrons, it becomes charged. A charged atom is called an **ion**. A positively charged ion has lost one or more electrons and is called a **cation**, while a negatively charged ion has gained one or more electrons and is called an **anion**. The charge or oxidation state on the ion is expressed as a *right-hand superscript* to the element symbol, e.g., $Ca^{2+}$.

G. Elements in a given vertical column are referred to as a group or family. The reason for this is that they have similar electron configurations and thus prefer to gain or lose the same number of electrons. This, in turn, helps predict their reactivity with other elements. There are eight major groups or families. For example, elements in the first two columns are known as alkali metals (starting with lithium and ending with francium) and alkaline earth metals (starting with beryllium and ending with radium), respectively. They prefer to lose one and two electrons from their outermost energy levels, respectively. They form cations, just as other metals do. Their ions are electron-deficient; hence, the number of protons is higher than the number of electrons. Alkali metals are said to have a valence or oxidation state of +1, or are univalent, while alkaline earth metals are said to have a valence or oxidation state of +2, or are divalent. Meanwhile, elements in the next to the last column, known as halogens, are nonmetals. They prefer to gain one electron and form an anion, and have a valence or oxidation state of −1. Both the tendency for metals to lose one or more electrons and the tendency for nonmetals to gain one or more electrons are explained by their electron configurations and energy stabilization rules, and account for the respective reactivities. There is a large array of metals in the middle of the periodic table known as the transition metals. Because of their complex electron configurations, they have multiple, energetically stable, positive valences or oxidation states and form a variety of interesting, industrially useful compounds. The seven elements that lie in between metals and nonmetals that form a descending step in the periodic table are known as metalloids. They have the ability to both gain and lose electrons, depending on their immediate chemical environment; their properties are often less predictable but equally interesting. Consult Table 1.1 below to see which elements form cations and which form anions, and which may form both.

In general, oxides of nonmetals, when dissolved in water, are acidic. Oxides of metals are alkaline or basic. Thus, substances such as calcium oxide and magnesium oxide are slightly alkaline in the presence of water. Substances such as carbon dioxide and sulfur dioxide, on the other hand, are slightly acidic. Normal rainwater, for example, has a pH of about 5.6 (slightly acidic) due to the presence of carbon dioxide in the atmosphere. For a distribution of important elements in the environment, see Appendix A1 through A5.

H. **Free Radicals** and **Ions** may occasionally be confused with each other, but they are quite different and have different stabilities. Ions are electrically charged atoms, having gained or lost one or more electrons, for reasons discussed in the preceding paragraph (G). They are quite stable in aqueous salt solutions. A list of common cations and anions is provided in Table 1.1.

## TABLE 1.1
### Names and Symbols of Simple Cations and Anions (Type I), Cations with Multiple Oxidation States (Type II) and Polyatomic/Oxo-Anions.

| Common Simple Cation and Anions | | | |
|---|---|---|---|
| Cation Symbol | Name | Anion Symbol | Name |
| $H^+$ | Hydrogen ion | $H^-$ | Hydride |
| $Li^+$ | Lithium ion | $F^-$ | Fluoride |
| $Na^+$ | Sodium ion | $Cl^-$ | Chloride |
| $K^+$ | Potassium ion | $Br^-$ | Bromide |
| $Cs^+$ | Cesium ion | $I^-$ | Iodide |
| $Be^{2+}$ | Beryllium ion | $O^{2-}$ | Oxide |
| $Mg^{2+}$ | Magnesium ion | $S^{2-}$ | Sulfide |
| $Ca^+$ | Calcium ion | $N^{3-}$ | Nitride |
| $Ba^{2+}$ | Barium ion | | |
| $Al^{3+}$ | Aluminum ion | | |
| $Ag^+$ | Silver ion | | |
| $Zn^{2+}$ | Zinc ion | | |
| $Cd^{2+}$ | Cadmium ion | | |

*(Continued)*

**TABLE 1.1** (*Continued*)

**Names and Symbols of Simple Cations and Anions (Type I), Cations with Multiple Oxidation States (Type II) and Polyatomic/Oxo-Anions.**

### Common Type II Cations

| Ion | Systematic Name | Older Name |
|---|---|---|
| $Fe^{3+}$ | Iron(III) | Ferric |
| $Fe^{2+}$ | Iron(II) | Ferrous |
| $Cu^{2+}$ | Copper(II) | Cupric |
| $Cu^+$ | Copper(I) | Cuprous |
| $Co^{3+}$ | Cobalt(III) | Cobaltic |
| $Co^{2+}$ | Cobalt(II) | Cobaltous |
| $Sn^{4+}$ | Tin(IV) | Stannic |
| $Sn^{2+}$ | Tin(II) | Stannous |
| $Pb^{4+}$ | Lead(IV) | Plumbic |
| $Pb^{2+}$ | Lead(II) | Plumbous |
| $Hg^{2+}$ | Mercury(II) | Mercuric |
| $Hg_2^{2+}$ | Mercury(I) | Mercurous |

### Names of Common Polyatomic Ions

| Ion | Name | Ion | Name |
|---|---|---|---|
| $NH_4^+$ | Ammonium ion | $CO_3^{2-}$ | Carbonate ion |
| $NO_2^-$ | Nitrite ion | $HCO_3^-$ | Hydrogen carbonate ion (bicarbonate ion commonly used) |
| $NO_3^-$ | Nitrate ion | $ClO^-$ | Hypochlorite ion |
| $SO_3^{2-}$ | Sulfite ion | $ClO_2^-$ | Chlorite ion |
| $SO_4^{2-}$ | Sulfate ion | $ClO_3^-$ | Chlorate ion |
| $HSO_4^-$ | Hydrogen sulfate ion (bisulfate ion commonly used) | $ClO_4^-$ | Perchlorate ion |
| $OH^-$ | Hydroxide ion | $CH_3COO^-$ or $C_2H_3O_2^-$ | Acetate ion |
| $CN^-$ | Cyanide ion | $MnO_4^-$ | Permanganate ion |
| $SCN^-$ | Thiocyanate ion | $(COO)_2^{2-}$ | Oxalate ion |
| $PO_4^{3-}$ | Phosphate ion | $Cr_2O_7^{2-}$ | Dichromate ion |
| $HPO_4^{2-}$ | Hydrogen phosphate ion | $CrO_4^{2-}$ | Chromate ion |
| $HPO_4^-$ | Dihydrogen phosphate ion | $O_2^{2-}$ | Peroxide ion |
| $AsO_4^{3-}$ | Arsenate ion | | |

*Note:*   The following halogens can form polyatomic ions in the same manner as chlorine: $BrO^-$, hypobromite ion; $IO^-$, hypoiodite ion; $BrO_2^-$, bromite ion; $IO_2^-$, iodite ion; $BrO_3^-$, bromate ion; $IO_3^-$, iodate ion; $BrO_4^-$, perbromate ion; $IO_4^-$, periodate ion.

Free radicals, on the other hand, concern unpaired electrons in electrically neutral molecules. Stable molecules are made when two atoms join together and form a bond in which two electrons are shared. This is a single bond and is consistent with the duet and octet rules for stable molecules. As will be seen in Chapter 3 in organic chemistry, double and triple bonds are also possible, in which four and six electrons are shared, respectively. In addition to bonding pairs, molecules may also have lone pairs of electrons, which do not participate in the bonding process, but serve to fulfill the octet rule for valence electrons. They also exist as pairs. However, occasionally, a molecule may exist in a meta-stable state having resonance structures (discussed in Chapter 3), where an atom has an unpaired electron. Such a molecule is referred to as a free radical, is highly reactive, and has a very short lifetime. One example of a free radical is the hydroxyl radical •OH [not to be confused with the hydroxide anion OH⁻]. The equation below shows the reaction of hydroxyl radical with a methane molecule to produce a methyl radical, another free radical:

$$\bullet OH_{(g)} + CH_{4(g)} \rightarrow \bullet CH_{3(g)} + H_2O_{(g)}$$

Free radicals are known to react with DNA in the human body, leading to its unraveling and breakdown in coding operations and proper replication. Consumption of antioxidants, such as L-ascorbic acid (vitamin C), which preferentially react with free radicals, is believed to reduce this problem.

**Example 1.1:** (A) Interpretating Symbols: Subscripts and Superscripts.
(B) Determining Protons, Neutrons and Electrons in an Isotope

Uranium, the element with atomic number $Z = 92$, has three naturally occurring isotopes: U-234, U-235, and U-238. Only U-235 is fissionable. When it combines with a halogen to form a compound, it commonly forms a cation with a +6 charge.

A. Using proper notation, write the symbol of the U-235 isotope, indicating its electronic charge (also known as the **valence** or **oxidation state**).
B. Determine the number of protons, neutrons, and electrons in one atom of this isotope.

*Solution*

A. The symbol, written with proper notation, is

$$^{235}_{92}U^{+6}$$

B. The atom has 92 protons (note left-hand subscript).

It must also have 92 electrons, understood, in its neutral (uncharged) atom. However, since this atom has a charge of +6 (note right-hand superscript), it has *lost* six electrons to become positively charged. Hence, it now has only 86 electrons.

To compute the number of neutrons in this atom, subtract the atomic number (left-hand subscript) from the mass number (left-hand superscript): $235 - 92 = 143$ neutrons.

### 1.2.2   Common Quantities and Units of Measurement

Important, measurable quantities or variables, along with their symbols and common units of measurement, are as follows:

- **Temperature**, *T*, in degrees centigrade or Celsius (°C), or Fahrenheit (°F), or in Kelvin units (K).
- **Mass**, *m*, in grams (g) or in kg (SI unit).
- **Amount, n, measured in moles**, one mole being $6.022 \times 10^{23}$ particles (e.g., atoms, molecules, or formula units); see Section 1.4.
- **Molar Mass**, *MM*, the mass in grams of $6.022 \times 10^{23}$ atoms of an element or molecules of a compound, i.e., one mole of a substance; see Section 1.4. The terms **molecular weight** and **formula weight** are often used interchangeably with molar mass.
- **Avogadro's Number** or **Constant:** $6.022 \times 10^{23}$ atoms, molecules, particles or items per mole.
- **Volume**, *V*, in cubic centimeters ($cm^3$) for a solid or liters (L) and milliliters (mL) for a liquid.
- **Pressure**, *P*, in atmospheres (atm), Pascals, torrs, or psi.
- **Density**, *d*, in grams per cubic centimeter ($g/cm^3$) for solids or grams per milliliter (g/mL) for a liquid.
- **Concentration** may be expressed in any one of several units, depending on the application; see Section 1.7.

Note that amounts of substances are expressed in *grams* (unit of mass) or *moles*. *Liters* is a unit of volume.

## 1.3   WRITING CHEMICAL FORMULAS AND NAMES

### 1.3.1   Inorganic Nomenclature and Formula Writing

There are two methods to name inorganic compounds and write their chemical formulas:

1. The crisscross method for ionic or ionically bonded compounds.
2. The Greek prefix method for covalent or covalently bonded compounds.

In order to decide which method is suitable, it is first necessary to decide what class of compound is being named: **ionic** or **covalent**. Recall that ionic compounds are made up of a metal and a nonmetal, where one or more electrons have been effectively transferred from one atom to another. For example, salts, such as sodium chloride or potassium sulfide, are ionic compounds. Covalent compounds are made up of two nonmetals; examples are carbon dioxide and phosphorous pentachloride. Covalent compounds are characterized by covalent bonds, in which electrons are shared between two atoms. If the sharing is unequal, the bond is said to be **polar covalent**, and one end of the molecule has a partial negative charge. Such molecules are referred to as **dipolar** and have dipole moments, which are quantitative measures of their polarity. [Higher-order polarities, such as quadrupoles (e.g. in $CO_2$) and hexadecapoles (e.g., in $SF_6$) do exist and can be calculated but are usually small and make an insignificant contribution to the net polarity.] If the sharing is equal, the bond is said to be nonpolar covalent; the molecule has no charge separation and hence no positive or negative end.

If the compound in question is ionic, use the crisscross method of nomenclature. Write the symbols of metal and nonmetal elements side by side (recall that a non-metal ion may be a single anion or a polyatomic oxo-anion, such as sulfate or nitrate), along with their respective valences or oxidation states as right-hand superscripts. Then crisscross the superscripts, that is, interchange them for the two elements, and write them as subscripts, omitting the plus or negative signs. If the two subscripts are the same, drop them both. In nomenclature, the name of the metal element is given first and that of the nonmetal element second, changing it to an "-ide" ending, such as chlor_ide_ or sulf_ide_, or using the name of the oxo-anion, such as sulfate or nitrate.

If a compound is covalent, the Greek prefix method is used. Here, knowledge of valences is unnecessary, but it is necessary to know the first 12 numbers (i.e., 1–12) in Greek:

• mono, di, tri, tetra, penta, hexa, hepta, octa, nona, deca, unideca, and dodeca.

The proper Greek prefix then precedes each element in the formula. As above, the name of the second element takes on the suffix "–ide."

It should be remembered that many compounds have chemical names as well as common names. A compound generally has only one correct chemical name but may have more than one common name. For example, the compound $Ca(OH)_2$ has the chemical name calcium hydroxide; it may also be referred to commonly as slaked lime or hydrated lime.

**Example 1.2:** (A–C) Writing Chemical Formulas of Ionic Compounds. (D–E) Writing Chemical Formulas of Covalent/Molecular Compounds

Write the correct chemical formula for each of the following compounds:

    A. Calcium chloride
    B. Aluminum sulfate
    C. Iron (III) phosphate
    D. Sulfur trioxide
    E. Diphosphorus pentoxide

### Solution

Compounds **A–C** are comprised of a metal and a nonmetal and are therefore ionic, so the crisscross method applies.

Compounds **D** and **E** are comprised of two nonmetals and are therefore covalent, so the Greek prefix method applies.

A. Calcium chloride: $Ca^{2+}$ and $Cl^-$ are the two ions (note that the 1 in front of the minus sign in the chloride ion has been omitted because 1's are understood and not written).

  Crisscrossing gives $CaCl_2$.

  Note that the subscript 1 next to Ca has been omitted because 1's are understood and are not written when specifying numbers of atoms.

B. Aluminum sulfate: $Al^{3+}$ and $(SO_4)^{2-}$ are the two ions.

  Crisscrossing gives $Al_2(SO_4)_3$.

  Note that parentheses are necessary to express the fact that the subscript 3 refers to and thus multiplies both the sulfur and oxygen atoms (subscripts) inside the parentheses.

C. Iron (III) phosphate: $Fe^{3+}$ and $(PO_4)^{3-}$ are the two ions.

  Note that the iron (III) notation refers to the $Fe^{3+}$ species, to be distinguished from the $Fe^{2+}$ species, written as iron (II). This is the modern way of distinguishing the ferric ion from the ferrous ion. This same system of notation, using Roman numerals in parentheses, is used for all transition metals with multiple oxidation states.

  Crisscrossing gives $Fe_3(PO_4)_3$.

  Note that fine-tuning is necessary in this case. Since both subscripts are 3, they are omitted. Furthermore, the parentheses then become unnecessary and are dropped.

  The fine-tuned answer is $FePO_4$.

D. Sulfur trioxide: S and O are the elements.

  Sulfur is "mono" or 1 (understood and not written), and oxygen is "tri" or 3. Hence, the formula is $SO_3$.

E. Diphosphorus pentoxide: P and O are the elements.

  Phosphorus is "di" or 2, and oxygen is "penta" or 5. Hence the chemical formula is $P_2O_5$.

**Example 1.3:** (A–C) Interpreting Subscripts in Chemical Formulas

Find the total number of atoms in one formula unit of compounds **A**, **B**, and **C** in Example 1.2.

*Solution*
In each case, the subscripts next to all the atoms in the formula unit are added together. Where parentheses are used, subscripts inside parentheses must first be multiplied by subscripts outside parentheses.

For $CaCl_2$: 1 Ca atom + 2 Cl atoms = 3 atoms total
For $Al_2(SO_4)_3$: 2 Al atoms + 3 S atoms + 12 O atoms = 17 atoms total
For $FePO_4$: 1 Fe atom + 1 P atom + 4 O atoms = 6 atoms total

*Note*: The percent oxygen (O) *by number* in $Al_2(SO_4)_3$ is $12/17 \times 100$ or 70.6%, while in $FePO_4$, it is $4/6 \times 100$ or 66.7%. Compare this calculation to the calculation for percent *by mass* in Example 1.6.

### 1.3.2  CALCULATING OXIDATION STATES FROM CHEMICAL FORMULAS

Occasionally, it may be necessary to determine the oxidation state or valence of an ion in a compound from its formula. This situation is often encountered with compounds or polyatomic complexes containing transition metal ions, which may have multiple oxidation states. Fortunately, many elements do have fixed oxidation states or valences as summarized in Table 1.2.

A simple algebraic equation solves the problem, as illustrated in Example 1.4.

**Example 1.4:** (A & B) Evaluating Oxidation States
in Molecules and Polyatomic Ions

Find the oxidation state or valence of each underlined atom.

   A. K$\underline{Mn}$O$_4$
   B. ($\underline{Cr}_2$O$_7$)$^{2-}$

*Solution*

Here, it is necessary to know the oxidation states assigned to common "fixed" elements. Consult Table 1.1. Some general guidelines are as follows:

   • Alkali metals are 1+.
   • Alkaline earth metals are 2+.
   • Oxygen is generally 2−.
   • Halides are generally 1−.

Set up a simple algebraic equation based on the principle that the sum of known charges of individual atoms in the formulas, multiplied by their respective subscripts, must total the net charge on the formula unit of the compound or polyatomic ion.

   A. For KMnO$_4$:

$$(1+)(1)+(x)(1)+(2-)(4)=0$$

Note that $x$ represents the unknown oxidation state of the Mn atom. The total is set to zero because the formula unit is electronically neutral or uncharged.

---

**TABLE 1.2**

**Oxidation State Assignments**

| | |
|---|---|
| F | −1 |
| Li, Na, K, Rb, Cs | +1 |
| Be, Mg, Ca, Sr, Ba | +2 |
| H | +1 |
| O | −2 |
| Cl, Br, I | −1 |

---

Solve for $x$: $x = +7$.
Thus, the oxidation state of Mn in $KMnO_4$ is $+7$.
B. For $(Cr_2O_7)^{2-}$:

$$(x)(2)+(2-)(7)=-2$$

Note that $x$ represents the unknown oxidation state of the Cr atom. The total is set to $-2$ because the net charge on the complex is $-2$.
Solve for $x$: $x = +6$.
Thus, the oxidation state of Cr in $(Cr_2O_7)^{2-}$ is $6+$.

## 1.4   CHEMICAL BONDING

Chemical bonding is a complex subject. Bonds that form between atoms to form compounds are referred to as intramolecular bonds, ionic, covalent, or metallic in nature. The discussion in the previous section was predicated on this type of bonding. A second, weaker category of bonding takes place between and among molecules, particularly organic molecules, and is referred to as intermolecular bonding. There are several types: hydrogen bonding, dipole–dipole interaction, London dispersion forces, to name just a few. This topic is covered in Section 1.9.6.

The subject of chemical bonding centers around the concept of the transfer or the sharing of electrons between atoms. Further, the sharing may be equal or unequal. The first type, i.e., transfer of electrons, is referred to as ionic bonding and is very common among salts (metal atoms bonding with nonmetal atoms), while the second type, i.e., sharing of electrons, is referred to as covalent bonding (and frequently occurs between two nonmetal atoms). If the sharing is purely equal, it is further referred to as nonpolar covalent bonding, while if the sharing is unequal, the bonding is referred to as polar covalent. In a nonpolar covalent bonded molecule, the molecule is essentially electrically neutral. In a polar covalent bonded molecule, one part or end of the molecule has a partial positive charge while the other end or part (where the electrons spend more time) has a partial negative charge. The degree to which a molecule is polar-covalent is determined by its dipole moment (or charge separation measured in units of debyes) and by its polarizability (a kind of shape or distortion factor). The type of bonding helps determine its physical and chemical properties, which in turn helps to determine its suitability in applications. For example, a long-chain hydrocarbon like octane present in gasoline is a nonpolar covalent molecule. Water, on the other hand, is a polar covalent molecule. A well-known principle of solubility in chemistry is "Like dissolves like." Thus, we can assume that octane will NOT dissolve in water, which is actually the case.

To help understand and predict these different types of bonding, it is helpful to show the following chart, Figure 1.1 Trends in Atomic Behavior, below.

It is also useful to define three related by different terms: ionization potential, electron affinity, and electronegativity (the latter not shown above).

**Ionization Potential/Energy:** The ionization potential or energy is the amount of energy required to remove the outermost electron from a ground state atom in its

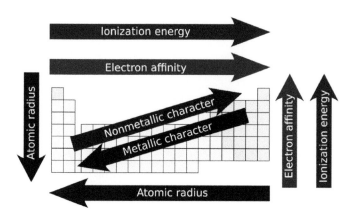

**FIGURE 1.1** Trends in atomic behavior.

gaseous state. This is known as the first ionization potential, usually measured in units of kilo-Joules/mole (kJ/mol) of atoms. This is usually relatively low for metal atoms which tend to lose electrons in a chemical reaction and form positive ions, and relatively high for nonmetallic atoms, which tend to attract electrons. Follow the red arrow in the Figure 1.1. For example, the lithium atom has a first ionization potential of 520 kJ/mol, meaning that this amount of energy would have to be applied to remove its outermost electron (i.e., its one 2S electron). This is a relatively low amount of energy when compared with a nonmetallic atom such fluorine, chlorine, or oxygen. If an atom has more than one electron, which can be removed, subsequent removal energies required are termed the "second" and "third" ionization potentials, which are normally, progressively higher.

**Electron Affinity:** Electron affinity is an analogous property for atoms that like to gain an electron in the gaseous state. This is particularly important for nonmetallic atoms. It is the energy change, i.e., the amount of energy given off, when an atom or ion accepts an electron and forms a negative ion. Follow the light brown arrow in the Figure 1.1. For example, a neutral fluorine atom will give off −328 kJ/mol when it accepts an electron and forms a fluoride ion. Fluorine has the highest electron affinity. The minus sign indicates energy is released or given off.

**Electronegativity:** Electronegativity is the ability of an atom in covalently bonded molecule to attract electrons to itself. It is the atom's pulling power to attract electrons to itself. The *difference* in pulling power or attracting electrons between the two atoms in a molecule is what determines its charge distribution and polarity. The scale of electronegativities runs from 0.0 to 4.0. Cesium and francium are alkali metals and have the lowest electronegativity values (often <1), while fluorine has the highest (3.98) followed by oxygen (3.44) and chlorine (3.16). The greater the difference in electronegativity values between two atoms bonded in a molecule, the greater the degree of polarity. All three of these quantities play a role in determining bond length, bond energy, and many other chemical and physical properties, and ultimately chemical reactivity.

## 1.5  CALCULATING MOLAR MASSES AND AVOGADRO'S PRINCIPLE

The atomic masses are given for each element in the periodic table in *amus* or atomic mass units. The mass number of a given isotope of an element is the sum of the number of protons and the number of neutrons in its nucleus. Recall that in Example 1.1, the mass numbers of the three isotopes of uranium were given as U-234, U-235, and U-238. However, this is not equivalent to the atomic weight. Most elements have more than one isotope, so the natural distribution of the isotopes of an element must also be taken into consideration. Using the carbon-12 isotope as the standard for mass, atomic masses can then be assigned to all the elements. For each element, this number must be a number that is averaged over all of its isotopes according to their relative percent natural abundance. The atomic weight of an element, then, is the average atomic mass of all of the element's naturally occurring isotopes. The molecular mass then becomes the sum of the atomic weights comprising the molecule, according to the number of each kind of atom occurring in the molecule. In other words, the molecular mass is the sum of the weights of the atoms represented in a molecular formula. Molecular masses (also called molecular weights) are the masses of molecules, which consist of essentially covalent compounds, while formula masses (also called formula weights) are the masses of formula units, which are essentially ionic compounds. The unit, in either case, is the amu, but often converted to the more useful grams/mole, which has the same numerical value.

To get from the microscopic level to the macroscopic scale, a quantity known as "the mole" (abbreviated mol) is defined. A mole of items, regardless of size, shape, or color, all of which are assumed to be identical, is equal to $6.022 \times 10^{23}$. This is a constant and is referred to as Avogadro's number. When one mole of atoms of an element is collected, the atoms have a collective mass (*molecular mass* for a molecule, or a *formula mass* for a formula unit) in **grams**. For comparison sake, one dollar bill or one small paper clip has a mass of about 1 gram. The more general term used to refer to either molecular mass or formula mass is **molar mass**.

**Example 1.5:** Compute the molar masses of (A) carbon dioxide and (B) calcium nitrate.

Compute the molar masses of carbon dioxide and calcium nitrate.

*Solution*

A. For $CO_2$: The atomic weight of carbon is 12.0 g/mol. The atomic weight of oxygen is 16.0 g/mol.
   Hence, the molar mass of carbon dioxide = 12.0 g/mol × 1 + 16.0 g/mol × 2 = 44.0 g/mol

B. For $Ca(NO_3)_2$:

   The atomic weight of calcium is 40.1 g/mol.
   The atomic weight of nitrogen is 14.0 g/mol.
   The atomic weight of oxygen is 16.0 g/mol.

Hence, the molar mass of calcium nitrate = 40.1 g/mol × 1 + 14.0 g/mol × 2 + 16.0 g/mol × 6 = 164 g/mol

## 1.6   DETERMINING THE PERCENT
## COMPOSITION OF A COMPOUND

It is often important to be able to compute the percent composition of a compound. This ability is valuable, for example, when comparing compounds such as fertilizers or mineral supplements to determine which fertilizer contains the most nitrogen per unit mass of fertilizer, or which mineral supplement contains the most calcium per unit mass of a tablet. This determination requires that the molar mass of each of the compounds in question be calculated first. It is assumed that the reader is familiar with calculating the molar mass of a compound from the atomic weights of the individual elements in the compound. Example 1.6 illustrates this method.

**Example 1.6:** (A & B) Determining Percent Composition
by Mass of Compounds

Determine the percent composition, by mass, of each of the following compounds:

   A. Calcium oxide, CaO (commonly known as lime or quicklime)
   B. Calcium carbonate, $CaCO_3$ (commonly known as limestone)

### Solution

   A. For CaO, the percent composition must be a two-part answer, because
   two elements make up CaO. Its molar mass is 56.1 g/mol. Thus:

$$\%Ca = \frac{\left(\text{atomic weight of Ca}\right)\left(\#\text{Ca atoms in 1 formula unit}\right)}{\text{molar mass of CaO}} \times 100$$

$$= \frac{\left(40.1 \text{ g Ca}\right)(1)}{56.1 \text{g CaO}} \times 100$$

$$= 71.5\%$$

$$\%O = \frac{\left(\text{atomic weight of O}\right)\left(\#\text{O atoms in 1 formula unit}\right)}{56.1 \text{g CaO}} \times 100$$

$$= \frac{\left(16.0 \text{g O}\right)(1)}{56.1 \text{g CaO}} \times 100$$

$$= 28.5\%$$

   B. For $CaCO_3$, the percent composition must be a three-part answer,
   because three elements make up $CaCO_3$. Its molar mass is 100.1 g/mol.
   In a similar fashion, then,

$$\%Ca = \frac{\left(40.1 \text{g Ca}\right)(1)}{100.1 \text{ g } CaCO_3} \times 100$$

$$= 40.0\% Ca$$

$$\%C = \frac{(12.0\,\mathrm{g\,C})(1)}{100.1\,\mathrm{g\,CaCO_3}} \times 100$$

$$= 12.0\%$$

$$\%O = \frac{(16.0\,\mathrm{g\,O})(3)}{100.1\,\mathrm{g\,CaCO_3}} \times 100$$

$$= 48.0\%$$

Note that CaO has a higher percent Ca content by mass than $CaCO_3$.

## 1.7   DETERMINING EMPIRICAL AND MOLECULAR FORMULAS FROM PERCENT COMPOSITION

It is frequently useful to determine the chemical formula of an unknown compound. One method is **combustion analysis**. If the unknown compound is known to contain at least carbon and hydrogen, heating a measured amount in the presence of excess oxygen, a process known as combustion, produces carbon dioxide and water, which can be collected and weighed individually. All of the carbon atoms are contained in the carbon dioxide, and all of the hydrogen atoms are contained in the water. Since 1 mole of carbon dioxide is equivalent to 1 mole of carbon atoms, the number of moles of carbon and hydrogen in the unknown compound can be determined. If oxygen is also present in the unknown, it is distributed over the carbon dioxide and water products; its mass can be found by difference and then converted to moles of oxygen in the original or unknown compound. The mass or moles of other elements, such as nitrogen and sulfur, which form known compounds with oxygen, can be determined in the same fashion.

Alternatively, another method is known as **elemental analysis**. A measured amount of the unknown substance is heated until it decomposes into its constituent elements, which are collected and analyzed individually. This method is called **pyrolysis** which is the intense heating of a compound or mixture in the absence of oxygen. The percent composition by mass of the compound is obtained. Example 1.7 illustrates how percent composition data can be used to deduce the chemical formula and ultimately the identity of an unknown substance.

**Example 1.7:** (A & B) Determining Empirical and Molecular Formulas

The percent composition by mass of an unknown compound is 40.9% carbon, 4.57% hydrogen, and 54.5% oxygen. By a separate analysis, its molecular weight or molar mass is found to be 176 g/mol. Find for the unknown compound:

   A. the empirical formula
   B. the molecular formula

## Solution

A. Recall that the empirical formula represents the smallest group or combination of atoms, in the proper ratio, of which the molecule is composed. Assume a 100-g sample to work with. Then, the given percentages can be translated directly into grams as follows:

$$40.9\,g\,C, 4.57\,g\,H, 54.5\,g\,O$$

Next, convert these masses to moles by dividing each one by its respective atomic mass:

$$40.9\,g\,C\frac{1.00\,mol\,C}{12.01\,g\,\,C} = 3.41\,mol\,C$$

$$4.57\,g\,H\frac{1.00\,mol\,H}{1.008\,g\,H} = 4.53\,mol\,H$$

$$54.5\,g\,\,O\frac{1.00\,mol\,O}{16.0\,g\,O} = 3.41\,mol\,O$$

At this point, the chemistry is done, and you could write the empirical formula in principle, using the calculated numbers as the subscripts: $C_{3.41}H_{4.53}O_{3.41}$.

Chemical formula rules stipulate, however, that these subscripts must not only be in the proper ratio but also must be whole numbers (integers). The problem now is to find a mathematical technique to maintain the ratio but change the subscripts into integers. One way is to divide through by the smallest number:

$$C_{3.41/3.41}H_{4.53/3.41}O_{3.41/3.41} = C_{1.00}H_{1.33}O_{1.00}.$$

Only the 1.33 subscript is not an integer. This can be corrected by multiplying through by the factor 3:

$$C_{3.00}H_{4.00}O_{3.00}$$

or simply written as $C_3H_4O_3$

This, then, is the empirical formula.

B. To find the molecular formula, divide the molar mass by the empirical mass. This ratio should always be a whole number (to three significant figures). Then, multiply each of the subscripts in the empirical formula by this factor to obtain the molecular formula.

The empirical mass of the above chemical formula ($C_3H_4O_3$) is calculated in the same way as a molar mass would be and equals 88.0 g/mol. Thus, the factor is:

$$\frac{Molar\,Mass}{Empirical\,Mass} = \frac{176\,g/mol}{88\,g/mol} = 2.00$$

Therefore, the correct molecular formula is $C_6H_8O_6$.

## 1.8   SOLUBILITY AND CONCENTRATION UNITS FOR AQUEOUS SOLUTIONS

Many chemical reactions and processes occur in the solution state where water is the solvent. Solubility of a substance is an important chemical property, since clearly, not all substances are soluble in water and can use water as a reaction medium. The term "soluble" deserves some clarification. When a salt such as sodium chloride is said to be soluble in water, it means that moderate, visible, or measurable amounts will dissolve. Even sodium chloride has an upper limit of about 35 g per 100 mL of water at room temperature. And similarly, even sand or trichloroethylene will dissolve to a limited extent in water, especially if the volume of water is sufficiently large. So, it is preferable to say that the salt calcium phosphate is slightly soluble instead of insoluble, though use of the latter term usually outnumbers the former. Solubility rules for most inorganic compounds in water are well known and are summarized in the table below. For organic compounds, whether solid or liquid, it is best to consult the Handbook of Chemistry and Physics or Wikipedia. Solvents other than water, such as acetone, alcohol, or hexane, are usually given or listed as alternatives (Table 1.3).

Several units exist in chemistry to quantify concentrations of aqueous solutions. The reason for the great variety is that different applications and chemical formulas require different units to keep these formulas mathematically manageable. The term "concentration" always implies the existence of a **solute**, present as a gas, liquid, or solid, dissolved in a suitable **solvent** to form a **solution**. Unless otherwise specified, the solvent for liquid solutions is understood to be water—hence the designation "aqueous solution," abbreviated as **(aq)**. The common definitions and concentration units are summarized in Figure 1.2.

---

### TABLE 1.3
### General Solubility Rules of Cations and Anions

| Soluble Compounds | Insoluble Exceptions |
|---|---|
| Compounds containing alkali metal ions ($Na^+$, $K^+$, $Rb^+$, $Cs^+$) and the ammonium ion ($NH_4^+$). | |
| Nitrates ($NO_3^-$), bicarbonates ($HCO_3^-$), Chlorates ($ClO_3^-$) | |
| Halides ($Cl^-$, $Br^-$. $I^-$) | Halides of $Ag^+$, $Hg_2^{2+}$, and $Pb^{2+}$ |
| Sulfates ($SO_4^{2-}$) | Sulfates of $Ag^+$, $Ca^{2+}$, $Sr^{2+}$, $Ba^{2+}$, $Hg_2^{2+}$, and $Pb^{2+}$ |

| Insoluble Compounds | Soluble Exceptions |
|---|---|
| Carbonates ($CO_3^{2-}$), phosphates ($PO_4^{3-}$), Sulfides ($S^{2-}$), chromates ($CrO_4^{2-}$) | Compounds containing alkali metal ions and the ammonium ion |
| Hydroxides ($OH^-$) | Compounds containing alkali metal ions and the $Ba^{2+}$ ion |

---

**Molarity, M** = moles of solute/liter of solution.
**Molality, m** = moles of a solute/kilogram of solvent.
**Mole Fraction, $X$** = number of moles of component $i$/total number of moles of all species.
**Percent by Mass/Mass** = mass of solute in grams/mass of solution in grams.
**Percent by Volume/Volume** = volume of solute in milliliters/total volume of solution in milliliters.

**FIGURE 1.2**   Common concentration units.

## Example 1.8: Preparing a Salt Solution Given Its Mass and Volume

A chemist wants to prepare 0.500 L of a 1.50 molar solution of sodium chloride. How does he/she do this?

### Solution

It is important to note that the unit of molarity provides the basis of making water-based chemical reactions and their stoichiometric calculations direct and much easier, since the unit of all stoichiometric coefficients in balanced chemical reactions is *the mole*. Thus, use is made of the simple but powerful relationship:

$$\text{Moles of Solute} = \text{Volume of Solution}(L) \times \text{Molarity}(M).$$

For example, the number of moles in 100 mL of a 2.5 M NaCl solution involved in a chemical reaction would be: moles NaCl = $(0.100\,L) \times (2.5\,M) = 0.250$ moles NaCl.

## Example 1.9: Interpreting Percent Concentration by Mass of a Salt Solution

Hospital saline solutions are often labeled as 1.25% KCl (aq) by mass. What does this mean, and how is this calculated?

### Solution

Using the definition of **Percent by Mass/Mass**, this means:

$$1.25\,g \text{ of } KCl(s)/100\,g \text{ of solution} \times 100 = 1.25\%$$

This solution is made by adding 1.25 g of KCl(s) to 98.75 g of water. Note that while masses of solutes and solvents are exactly additive, volumes of *different* solutions (with different densities) are not.

## Example 1.10: Interpreting Percent Concentration by Volume of a Solution

A wine bottle has a label that reads 11.5% alcohol (i.e., ethanol) by volume. What does this mean, and how is this calculated?

### Solution

Using the last definition above, this means:

$$11.5\,mL\,C_2H_5OH\,/\,100\,mL \text{ of wine solution} \times 100 = 11.5\%$$

Since most wine bottles contain 750 mL of fluid, this would mean there is 86.25 mL of pure ethanol in a bottle of wine.

Conversion from one concentration unit to another may occasionally be necessary or useful. The "factor label method" is the most efficient way to handle this. Note the following example.

**Example 1.11:** Converting Density of a Solution to Molarity

Find the concentration of sulfuric acid solution, $H_2SO_4$ (aq), used in car batteries, in molarity (M). The density of battery acid is about 1.265 g soln/mL soln (while pure sulfuric acid stock solution is 1.800 g/mL, for comparison), and it is 35.0% by mass $H_2SO_4$.

*Solution*

Start with the density figure and then proceed as follows:

$$1.265\,\text{g soln/mL soln}\left(\frac{1000\,\text{mL soln}}{1.00\,\text{L soln}}\right)\left(\frac{35.0\,\text{g}\,H_2\,SO_4}{100\,\text{g soln}}\right)\left(\frac{1.00\,\text{mol}\,H_2\,SO_4}{98.0\,\text{g}\,H_2\,SO_4}\right)$$

$$= 4.52\,\frac{\text{mol}}{\text{L}} = 4.52\,\text{M}$$

Note the systematic cancellation of units from left to right to get to moles $H_2SO_4$ per liter (L) of solution on the right, i.e., the unit of molarity, M. Other unit-to-unit conversions are performed in much the same way.

## 1.9   PROPERTIES OF GASES AND THE GAS LAWS

### 1.9.1   THE KINETIC-MOLECULAR THEORY OF GASES

The fact that all gases behave similarly with change in temperature or pressure led to the kinetic-molecular theory or model. It can be summarized as follows:

- All matter in the gas phase is composed of discrete, point particles called molecules.
- In the gaseous state, molecules are relatively far apart.
- The molecules of all substances in the gaseous state are in continuous, rapid, straight-line motion. This motion is in three dimensions and is often called translation.
- This continuous motion is a measure of the kinetic energy of the system.
- In addition to translation, molecules may also rotate and vibrate, depending upon the type of external energy source they are subjected to. These are known as internal modes of energy and separate from its kinetic energy.
- Collisions between molecules are assumed to be perfectly elastic. This means that when molecules collide with one another or with the walls of a container, they rebound without any loss of kinetic energy.

- The average kinetic energy of the molecules is directly proportional to its temperature in Kelvins. All gasses at the same temperature will have the same kinetic energy. Particles in a gas are assumed to exert no attractive or repulsive forces.

This leads to gases having the following general properties:

1. Gases exert pressure, which can be measured in any number of units, e.g., pounds per square inch, atmospheres, Pascals (Newtons per square meter), and Torrs.
2. Gases are highly compressible (unlike liquids and solids).
3. Gases diffuse easily.
4. Gases expand upon heating, provided the pressure remains constant.
5. The pressure exerted by a gas increases with temperature, provided the volume is held constant.

What is an ideal gas?

The model or assumptions for an ideal gas is twofold. First, it is assumed that the molecules consist of perfect spheres, and they take up no room or volume themselves when compared with the volume of the container they are in (sometimes referred to as excluded volume). The second is that the individual gas molecules do not see or attract one another. This model then leads to the **Ideal Gas Law**, which states:

$$PV = nRT \qquad (1.1)$$

where $P$ is the pressure of the gas system, $V$ is the volume of the container of the system, $T$ is the temperature of the gas (assumed to be the same as the temperature of the system container), $n$ is moles of gas, and $R$ is the universal gas constant, given in several different units in Table 1.4.

**TABLE 1.4**

**van der Waal Constants for Common Gases and Volatile Liquids**

| Gas | A (L²bar/mol²) | B (L/mol) |
|---|---|---|
| Argon | 1.355 | 0.03201 |
| Butane | 14.66 | 0.1226 |
| Carbon dioxide | 3.640 | 0.04267 |
| Carbon monoxide | 1.505 | 0.03985 |
| Ethanol | 12.18 | 0.08407 |
| Nitric oxide | 1.358 | 0.02789 |
| Nitrogen | 1.370 | 0.0387 |
| Nitrogen dioxide | 5.354 | 0.04424 |
| Nitrous oxide | 3.832 | 0.04415 |
| Oxygen | 1.382 | 0.03186 |
| Radon | 6.601 | 0.06239 |
| Sulfur dioxide | 6.803 | 0.05636 |

This law or formula is easy to work with and accurately predicts the behavior of all the ideal gases (helium, argon, etc.) as well as some commonly encountered gases such as nitrogen and oxygen, the two chief components of air, at or near room temperature and pressure.

However, under conditions of high pressure or low temperature, gases do not behave ideally. There are two reasons for this deviation from ideal behavior. First, the molecules themselves have a definite volume and size and so do occupy a significant fraction of the volume of the container under high pressure. This causes the volume of the gas to be greater than that *calculated* for an ideal gas. Hence, a subtractive correction factor—"B"—is necessary. B is a measure of the excluded volume of one mole of particles. Second, molecules are brought closer together, and hence attract one another more strongly especially under conditions of high pressure or low temperature. Hence, an additive correction factor—"A"—is necessary. A is a measure of the attractive forces between particles. The van der Waals equation, Equation 1.2, represents *real* gas behavior under either of these two conditions and especially when both conditions prevail. It is mathematically more complicated and therefore more difficult to work with. The constants "A" and "B" may be looked up by gas identity in a table in a handbook. An abridged table of values is given in Table 1.4.

$$\left[ P + n^2 A / V^2 \right] \left[ V - nB \right] = nRT \tag{1.2}$$

Besides the ideal gas law and the van de Waals models, there are no fewer than ten other models for predicting the behavior of gases under real conditions. For the most part, they are far more mathematically complex. However, these two models and their corresponding equations accurately predict the behavior of almost all gases and vapors under most operating conditions and still retain relative mathematical simplicity (Table 1.5).

### 1.9.2 The General Gas Law

**Boyle's Law** (the pressure of a gas is inversely proportional to its volume) and **Charles's Law** (the volume of a gas is directly proportional to its temperature) may be generally combined into a more useful form known as the **General Gas Law**, expressed as

$$\frac{P_1 V_1}{T_1} = \frac{P_2 V_2}{T_2} \tag{1.3}$$

**TABLE 1.5**

**The Universal Gas Constant "R" in Various Units**

0.082057 L·atm/mol·K

1.987 cal/mol·K

8.3145 J/mol·K

8.3145 m³·Pa/mol·K

62.364 L·Torr/mol·K

It is understood that his law is valid only for a closed system in which the total number of gas molecules and hence the mass are constant. The general gas law can be applied to an individual gas (e.g., nitrogen or carbon dioxide) or a mixture of gases (e.g., air). It is assumed that the gas is ideal, that is, follows the ideal gas law (see next section) as its equation of state.

Pressures $P$ and volumes $V$ may be in any units as long as they are consistent. Temperature must be absolute and thus in Kelvins. The subscripts 1 and 2 represent the initial and final states or conditions, respectively, for each quantity.

**Example 1.12:** Using the General Gas Law

A fixed quantity of gas occupies a volume of 2.0 L at a temperature of 20°C and a pressure of 1.0 atm. Find the volume that this gas would occupy at 40°C and 1.75 atm pressure.

*Solution*

The general gas law (Equation 1.3) is applied. This is a problem in which the volume of a gas must be calculated at a new or different temperature and pressure. The data given are as follows:

$$P_1 = 1.0\,\text{atm} \quad P_2 = 1.75\,\text{atm}$$

$$T_1 = 20°C + 273\,\text{K} = 293\,\text{K}$$

$$T_2 = 40°C + 273\,\text{K} = 313\,\text{K}$$

$$V_1 = 2.0\,\text{L} \quad V_2 = ?$$

Substitute these values in the general gas law and solve for $V_2$.

$$V_2 = \frac{P_1 V_1 T_2}{P_2 T_1} = \frac{(1.0\,\text{atm})(2.0\,\text{L})(313\,\text{K})}{(1.75\,\text{atm})(293\,\text{K})}$$

$$= 1.22\,\text{L}$$

## 1.9.3 THE IDEAL GAS LAW

The **ideal gas law**, known as the **equation of state** for ideal gases, is a very useful and powerful problem-solving tool. It relates the pressure, volume, and temperature of a quantity of a gas. An **ideal gas** is a gas in which every molecule behaves independently of every other molecule (there is an absence of any intermolecular forces) and has no excluded volume. It is represented by the ideal gas law stated earlier:

$$PV = nRT \tag{1.4}$$

where

$P$ = the pressure in atmospheres, torr, millimeters of mercury, pounds per square inch, pascals, etc.

$V$ = the volume in liters, cubic centimeters, etc.

$n$ = the number of moles of gas = mass of gas in grams/molar mass in grams per mole = $\dfrac{m}{MM}$

$T$ = the absolute temperature in Kelvin units

$R$ = the universal gas constant in units consistent with the above values, i.e., check Table 1.5.

The ideal gas law contains four variables and the constant R. Given any three, the fourth can be found. In addition, this law can be used to find the molar mass or molecular weight if the mass of a gas is given.

Note that the ideal gas law is valid for a single, individual gas. If a mixture of gases is present, the ideal gas law is valid for each individual gas. In other words, a separate calculation using this law can be made for each gas present. The pressure of each gas thus calculated represents a partial pressure in the mixture. The total pressure can then be calculated according to Dalton's Law (see the next section). However, to reduce calculation steps, this law can also be used to represent the total gas mixture as well.

### Example 1.13: Using the Ideal Gas Law

Exactly 5.75 g of an unknown gas occupies 3.49 L at a temperature of 50°C and a pressure of 0.94 atm. Find the molar mass or molecular weight of the gas.

### Solution

The ideal gas law (Equation 1.1) has many applications. One of them is to determine the molar mass of a gas.

$$PV = nRT = \left( \frac{m}{MM} \right) RT$$

In this case, solve for MM.

The data given in the problem are as follows:

$P = 0.94$ atm
$R = 0.0821$ L·atm/K·mol
$V = 3.40$ L
$T = 273 + 50°C = 323$ K
$m = 5.75$ g

Substitute this information in the above equation, and solve or *MM*:

$$MM = \frac{mRT}{PV}$$

$$= \frac{(5.75\,g)(0.0821\,L \cdot atm\,/\,K \cdot mole)(323\,K)}{(0.94\,atm)(3.40\,L)}$$

$$= 47.7\,g/mol$$

### 1.9.4   DALTON'S LAW OF PARTIAL PRESSURES

**Dalton's Law** states: The total pressure exerted by a mixture of gases is equal to the sum of the partial pressures of all the gases in the mixture. Each partial pressure is the pressure that the gas would exert if the other gases were not present.

$$P_{total} = \sum P_i = P_1 + P_2 + P_3 + \ldots + P_n \qquad (1.5)$$

This law is particularly useful when gases are collected in vessels above the surface of an aqueous or nonaqueous solution. In an aqueous solution, the partial pressure of water vapor may be subtracted from the total pressure to help determine the partial pressures of the other gases or vapors present. The percent composition of a gaseous mixture may then be calculated, based on Dalton's Law. To calculate the composition of the *solution phase* (e.g., in mole percent), **Raoult's Law** (see Equation 3.11) must be used.

### 1.9.5   GRAHAM'S LAW OF EFFUSION

Recall that kinetic molecular theory states that the average speed of molecules in motion can be approximated by the **root-mean-square speed, U$_{rms}$**:

$$U_{rms} = \sqrt{\frac{3RT}{MM}} \qquad (1.6)$$

**Diffusion** is the migration and mixing of molecules of different substances as a result of a concentration gradient across a fixed space and random molecular motion. **Effusion** is the escape of gas molecules of a single substance through a tiny orifice (pinhole) of a vessel holding the gas. **Graham's Law of Effusion** states: The rates of effusion of two different gases escaping, A and B, are inversely proportional to the square roots of their molar masses:

$$\frac{\text{Effusion Rate}_A}{\text{Effusion Rate}_B} = \sqrt{\frac{MM_B}{MM_A}} \qquad (1.7)$$

Note that, at a fixed temperature, the *effusion time* is inversely proportional to the effusion rate, while the *mean distance traveled*, as well as the *amount of gas effused*, is directly proportional to the effusion rate.

### 1.9.6  INTERMOLECULAR FORCES

Chemical bonds between two or more atoms in a molecule are referred to as **intramolecular forces**. They are generally categorized as ionic, covalent, or metallic. These are the bonds that are broken and reformed during a chemical reaction. Significantly lower in strength are **intermolecular forces**. Intermolecular forces are generally attractive forces that exist between and among molecules of all shapes, sizes, and masses. They exist in the gaseous, liquid, and solid phases, as well as in the solution phase, where water is acting as the solvent to hydrate the solute.

They are important to understand because they are directly related to macroscopic properties such as melting point, boiling point, vapor pressure and volatility, and the energy needed to overcome forces of attraction between molecules in changes of state. Just as important, they also help determine the solubility of gases, liquids, and solids in various solvents, i.e., whether two substances are soluble or miscible in each other, and in part explain or reflect the "like dissolves like" principle. They are also critical in determining the structure of biologically active molecules such as DNA and proteins. Finally, they help explain sources of failure in paints, coatings, and adhesives in fields such as construction engineering and consumer product malfunction.

The categories of intermolecular, attractive forces for neutral or uncharged molecules in order of increasing strength are as follows:

- London dispersion forces—exhibited by all molecules
- Polar forces or interactions—exhibited by all asymmetric molecules
- Hydrogen bonding—exhibited by molecules containing O-H, N-H, or F-H bonds.

More generally, however, these forces of attraction, including charged and uncharged species, can be usefully organized and expanded, again in order of increasing strength, as follows:

- London dispersion forces—important in nonpolar substances (substances having no permanent dipoles).
- Ion–dipole forces—important in salts (consisting of ions) dissolved in water (aqueous solutions), also referred to as "hydration forces."
- Dipole–dipole forces—important in polar, covalent substances, which have permanent dipoles.
- Hydrogen bonding, important in O-H, N-H, and F-H interactions.
- Ion–ion forces—important for ions only, usually present in aqueous solutions.

The term "van der Waals forces" refers to dipole-induced dipole interactions and is a special case of the more general term "London dispersion forces."

Note that the distance between the oxygen and hydrogen atoms within the water molecule is about 100 pm (where *pm* represents *picometer*, which is $1 \times 10^{-12}$ m). This is an *intramolecular* distance. In contrast, the distance between the oxygen atom of one water molecule and the hydrogen atom of a neighboring water molecule is about 180 pm. This is the *intermolecular* distance, typical in a hydrogen bond, and is almost double the intramolecular distance.

For example, polar substances such as acetone ($C_3H_6O$) will dissolve in other polar substances such as methyl chloride ($CH_3Cl$) but not in carbon tetrachloride ($CCl_4$), a nonpolar substance. Some polar substances, such as ethanol ($C_2H_3OH$), will dissolve in water in all proportions, i.e., are completely miscible, because of hydrogen bonding between the hydrogen atom of the ethanol molecule and the oxygen atom of the water molecule. The same is true of ammonia and hydrofluoric acid, where hydrogen bonding occurs. Note that dimethyl ether ($C_2OH_6$), an isomer of ethanol, does not undergo hydrogen bonding with water because of its different molecular structure, i.e., an ether versus an alcohol (see Chapter 3, Table 3.6—ethanol vs. dimethyl ether). Hexane ($C_6H_{14}$) will dissolve in octane ($C_6H_{18}$) because both substances are characterized by nonpolar bonds. Naturally, size (length) and shape (branching of the molecule) also play a role in solubility considerations, and it is often difficult to determine which factor is more important in predicting solubility between two substances.

The greater the forces of attractions between molecules in a liquid, the greater the energy that must be supplied to separate them. Hydrogen bonding is a key reason why low-molecular-weight alcohols have much higher than expected boiling points, i.e., heat of vaporization, in comparison to nonpolar hydrocarbons like hexane.

## Two Special Cases

The following two examples show the importance and relevance of intermolecular forces.

### Case 1: Bubbles Can Disappear

An interesting problem presented itself to a mechanical engineer who was trying to spread a thin film of adhesive onto a flat plastic surface to bond with another flat surface. The adhesive used was a hexane-based polymer, and a pressurized air gun was used to spread it as evenly as possible through a spreading nozzle. However, each time it was spread, tiny bubbles randomly appeared in the adhesive, preventing a tight or perfect seal. The engineer eventually realized that this was due to the fact that air, which is composed of nitrogen and oxygen, has a finite solubility in hexane, since both are nonpolar substances. Recalling the nature of intermolecular forces and solubility, a rough rule of thumb is that "like dissolves like." When helium, an inert gas, was substituted for air in the delivery mechanism, the bubbles disappeared, and the seal was perfect. Helium has virtually no solubility in hexane.

### Case 2: Supercritical Fluids "To the Rescue"

Supercritical fluids are versatile solvents because their properties vary significantly with changing temperature and pressure. A supercritical fluid can be made to dissolve or extract one component of a mixture without disturbing the others, simply by altering its temperature or pressure or both.

Thus, it can be used to remove environmental contaminants such as diesel fuel, PCBs (polychlorinated biphenyls), or OCP (organochlorine pesticides) from soil. In this case, carbon dioxide subjected to pressures as high as 400 atmospheres works nicely to extract a variety of congeners of PCBs or OCPs from soil. Carbon dioxide is now also the solvent of choice in the dry cleaning industry, having replaced carbon tetrachloride banned since 1978 and less toxic chlorinated hydrocarbons used until the late 1990s.

# 2 Simple Categories of Inorganic Chemical Reactions

## 2.1 FIVE GENERAL CATEGORIES OF CHEMICAL REACTIONS

The purpose of categorizing a chemical reaction is to help predict the products of the reaction. In general, a chemical reaction is a process of molecular rearrangements, in which atoms change partners. Atoms are neither destroyed nor created. Total mass is conserved, that is, the mass of all of the reactants before the reaction must equal the mass of all of the products after the reaction, although the physical or chemical states of individual substances may change. This mass-conservation requirement explains why all chemical reactions must be properly balanced to be quantitatively valid and useful. Energy changes also accompany reactions—energy is either liberated or absorbed—since existing chemical bonds are broken and new bonds are formed.

With the letters A, B, C, and D used to represent simple elements or polyatomic ions, and combination of two capital letters used to designate compounds, the five categories of reactions are as follows:

- Combination or Synthesis
  $$A + B \rightarrow AB$$
- Decomposition
  $$AB \rightarrow A + B$$
- Single Replacement/Displacement
  $$A + BC \rightarrow AC + B$$
- Double Replacement/Displacement
  $$AB + CD \rightarrow AD + CB$$
- Combustion (complete or 100%)
  $$(CH)_x + O_2 \rightarrow H_2O + CO_2$$

The notation (g), (l), (s), or (aq) immediately following a reactant or product is often employed to designate the physical state of the substance.

Besides these five general categories, there are three other categories of reactions that cut across these five. They can be viewed as subcategories of the above reactions, but are encountered so frequently that they are often designated as separate classes of chemical reactions. They are referred to as follows: (1) oxidation–reduction reactions, (2) acid–base reactions, and (3) equilibrium reactions.

In oxidation–reduction reactions, also called "red-ox reactions," electrons are transferred between atoms, either individually or within molecules. In acid–base

DOI: 10.1201/9781003092759-2

reactions, protons, in the form of hydrogen ions, are transferred between molecules. Equilibrium reactions, commonly encountered in organic chemistry, are interesting in that they are reactions that may proceed both forward and backward at the same time and are thus considered dynamic (not static) in nature. Note that it is possible for a reaction to fall into two or more categories or classes at the same time. For example, an acid–base reaction may also be an equilibrium reaction AND an example of a double replacement/displacement reaction.

Each of these classes of chemical reactions will be explained in more detail in subsequent chapters. Oxidation–reduction reactions are introduced below and discussed again in Chapter 7. Acid–base reactions will be introduced and discussed in detail in Chapter 3, while equilibrium reactions will be introduced and discussed in Chapter 4.

## 2.2   OXIDATION–REDUCTION OR RED-OX REACTIONS

As mentioned in the preceding section, a class of reactions that cuts across all other reaction categories consists of **oxidation–reduction reactions**. Oxidation–reduction reactions ("red-ox" reactions) are reactions in which one substance is oxidized while another is simultaneously reduced. The processes of oxidation and reduction can be defined as follows. Oxidation is the loss of electrons, while reduction is the gain of electrons. However, diagnostically speaking, a substance is oxidized or reduced, respectively, if any *one* of the following conditions is met:

**Oxidation**

- The substance loses electrons.
- The substance gains oxygen atoms.
- The substance loses hydrogen atoms.

**Reduction**

- The substance gains electrons.
- The substance loses oxygen atoms.
- The substance gains hydrogen atoms.

Red-Ox reactions are actually formed by the addition of two half-reactions—an oxidation reaction and a reduction reaction. It is impossible to have an oxidation without reduction or reduction without oxidation in a red-ox reaction. If one substance is oxidized, then another must be reduced, and vice versa, in the total or net reaction.

For example, each of the following shows an *oxidation reaction:*

$$Zn \rightarrow Zn^2 + 2e^-$$

$$\left[ Zn \text{ has lost 2 electrons.} \right]$$

(2.1)

$$C \rightarrow O_2 \rightarrow CO_2$$

$$\left[ C \text{ has gained } 2O \text{ atoms.} \right]$$

(2.2)

And each of the following shows a *reduction reaction*:

$$Cu^2 + 2e^- \rightarrow Cu$$

$$\left[ Cu^{2+} \text{ has gained 2 electrons.} \right] \tag{2.3}$$

$$C + 2H_2 \rightarrow CH_4$$

$$\left[ C \text{ has gained 4 H atoms.} \right] \tag{2.4}$$

Each half-reaction also has an associated electromotive potential, measured in electron volts (eV) and found in electromotive potential tables. This topic is discussed more fully in Chapter 2, Section 2.8.

An interesting case of red-ox reactions is **disproportionation**. In this type of reaction, a single substance undergoes both oxidation and reduction at the same time. An example of this is the disproportionation of $Cl_2$ gas placed in water as a disinfecting agent. The reaction proceeds as follows:

$$Cl_2(g) + H_2O(l) \rightarrow H^+(aq) + HOCl(aq) \tag{2.6}$$

Here, the initially neutral diatomic chlorine molecule has undergone reduction both to the chloride anion, $Cl^-$, and simultaneously to the chlorine cation, $(Cl^+)$, in the HOCl molecule, known as hypochlorous acid. The actual disinfection of bacteria is carried out by the $Cl^+$ ion present in HOCl, which is a strong oxidizing agent.

## 2.3 WRITING AND BALANCING CHEMICAL REACTIONS

A chemical reaction consists of a reactant side (the left) and a product side (the right) separated by an arrow, which indicates *yields* or *produces*. Reactants and products are present as elements and/or compounds and are represented by appropriate symbols, as discussed in Section 3.1. Reactions must be balanced before they can be used in calculations to provide quantitative information. Again, a symbol placed immediately after the element or compound designates its chemical state: (g) for gas or vapor; (l) for liquid; (s) for solid, powder, precipitate, or crystal; and (aq) for solution, where water is understood to be the solvent for the indicated solute.

The simplest balancing method is called *balancing by inspection*. Although this method is often interpreted to mean balancing by trial and error, three simple rules regarding the order of balancing should be followed:

1. Balance metal atoms or atoms of any element present in the greatest number first. Greatest number means the largest subscript next to the atom.
2. Balance nonmetal atoms second, in descending order of magnitudes by checking subscripts, as in step #1 above.
3. Balance hydrogen atoms next to last and then oxygen atoms last.

Balancing means inserting integers, known as stoichiometric coefficients, in front of elements or compounds to ensure the same number of like atoms on both sides of the reaction. Balancing requires keeping track of every kind of atom, not molecule, that appears in the reaction. Compare atoms of a particular species on the left (reactant) side of the arrow with atoms of the same species on the right (product) side of the arrow. Use of fractional coefficients, such as ½ or 1/3, is permitted as a temporary way of balancing atoms, particularly hydrogen and oxygen atoms. Be sure to multiply each coefficient by 2 or 3, respectively, to eliminate the presence of fractional coefficients in the final reaction. Subscripts of atoms in compounds are fixed by nature and cannot be altered to achieve balancing.

The following are examples of **balanced reactions**:

A. $2Al(s) + 6HCl(aq) \rightarrow 2AlCl_3(aq) + 3H_2(g)$

B. $3CaCl_2(aq) + 2Na_3PO_4(aq) \rightarrow Ca_3(PO_4)_2(s) + 6NaCl(aq)$

C. $2C_4H_{10}(g) + 13O_2(g) \rightarrow 8CO_2(g) + 10H_2O(g)$

D. $CaCO_3(s) \rightarrow CaO(s) + CO_2(g)$

The following points regarding these four reactions are noteworthy:

1.
   A.  is an example of a single-replacement reaction, since a single element is reacting with a compound.
   B.  is an example of a double-replacement reaction, since two compounds are reacting with each other.
   C.  is an example of a combustion reaction, since a hydrocarbon is reacting with oxygen to produce carbon dioxide and water.
   D.  is an example of a decomposition reaction, since a single compound is decomposing.
2. Only (C) is an example of a **homogeneous reaction**, since all of the reactants and products are present in the same physical state—gaseous. The other three are **heterogeneous reactions**, since the reactants and products are present in more than one physical state.
3. (A) and (C) are also examples of oxidation–reduction reactions. In (A), Al is oxidized, while HCl is reduced; HCl is the **oxidizing agent**, while Al is the reducing agent. In (C), $C_4H_{10}$ is oxidized while $O_2$ is reduced; $O_2$ is the oxidizing agent, while $C_4H_{10}$ is the **reducing agent**.
4. There are two, and only two, levels of interpretation of the stoichiometric coefficients that balance these reactions. Consider (C). On the microscopic or invisible level, the reaction states that 2 molecules of $C_4H_{10}$ react with 13 molecules of $O_2$ to produce 8 molecules of $CO_2$ and 10 molecules of $H_2O$. On the macroscopic or visible level, the reaction states that 2 moles of $C_4H_{10}$ [2 times the molar mass of $C_4H_{10} = (2\,mol)(58\ g/mol) = \mathbf{116\ g}$] react with 13 moles of $O_2$ [13 times the molar mass of oxygen $= (13\,mol)$ $(32\ g/mol) = \mathbf{416\ g}$] to produce 8 moles of $CO_2$ [8 times the molar mass of

$CO_2 = (8\,\text{mol})(44\text{ g/mol}) = \mathbf{352\ g}$] and 10 moles of $H_2O$ [10 times the molar mass of $H_2O = (10\,\text{mol})(18\text{ g/mol}) = \mathbf{180\ g}$]. In other words, the only two correct units of stoichiometric coefficients are molecules and moles. The unit grams can be obtained by the use of molar masses, as shown, but are not directly readable.

5. Note, from point #4 above, that in (C), total mass is automatically conserved. The total mass of reactants is $116 + 416\text{ g} = 532\text{ g}$. The total mass of products is $352 + 180\text{ g} = 532\text{ g}$. The conservation of mass principle requires that this equality always occurs.

## 2.4   SIMPLE STOICHIOMETRY

The terms "stoichiometry" refers to the mass relationships between two reactants, two products, or, more commonly, a reactant and a product. **Stoichiometry** depends on a balanced reaction, which is essential in predicting the amount of product generated by a given amount of reactant or, conversely, the amount of reactant required to generate a given amount of product. The method recommended is the **Factor-Label Method**, illustrated in Example 2.1.

**Example 2.1:** Stoichiometry Calculation of a Combustion Reaction

Consider reaction (C) in the previous section, the complete combustion of butane with oxygen to produce carbon dioxide and water. Calculate the mass of carbon dioxide that can be produced from 816 g of butane, assuming oxygen is in abundant or unlimited supply.

*Solution*

Since the reaction is balanced, the solution may proceed by setting up three conversion factors, written in parentheses and arranged from left to right, multiplying the given mass of butane. Next, draw a connecting "tie" line between the **2** in front of the butane and the **8** in front of the carbon dioxide, since these are the only two substances of interest. This will form the basis of the middle conversion factor below. The first and third conversion factors are the molar masses of butane and carbon dioxide, respectively. Thus:

$$2C_4H_{10}(g) + 13O_2(g) \rightarrow 8CO_2(g) + 10H_2O(g)$$

$$816\text{ g }C_4H_{10} \times \left(\frac{1.0\text{ mol }C_4H_{10}}{58.0\text{ g }C_4H_{10}}\right) \times \left(\frac{8.0\text{ mol }CO_2}{2.0\text{ mol }C_4H_{10}}\right) \times \left(\frac{44.0\text{ g }CO_2}{1.0\text{ mol }CO_2}\right) = 2,480\text{ g }CO_2$$

The mass of water can also be determined in a separate but similar calculation. The "tie" line would be drawn between the **2** in front of butane and the **10** in front of water. And the molar mass of water is 18.0 g/mol, instead or 44.0 g/mol for carbon dioxide. Thus, the mass of water could be computed to be 1266 g, or 1270 g rounded to three significant figures.

Note that the total mass of all (i.e., both) products would be the sum of the mass of carbon dioxide (2480 g) and the mass of water (1270 g), which equals 3750 g. Further, the principle of the conservation of mass stipulates that the total mass of all reactants must equal the total mass of all products, 3750 g must also represent the total mass of all reactants. It follows that since the original mass of butane was given as 816 g, the mass of oxygen that reacted must be 3750–816 g or 2934 g, or 2930 g to three significant figures.

Now consider reaction (B) from the previous section. It is one of thousands of reactions that occur in water. The reaction shows the precipitation of calcium phosphate when solutions of sodium phosphate and calcium chloride are mixed. This reaction can be written in three different ways. In its current form, it is written as a "molecular" reaction. It can also be rewritten in a more useful form called the "**total ionic**" equation, as follows:

$$3Ca^{2+}(aq) + 6Cl^-(aq) + 6Na^+(aq) + 2PO_4^{3-}(aq) \; Ca_3(PO_4)_2(s)$$
$$+ 6Na^+(aq) + 6Cl^-(aq)$$

The benefit of this form is that it shows all of the ions that are present in the aqueous state (i.e., hydrated by water), separate from those that are present in the solid state (or gaseous in another reaction), and gives a more accurate, detailed representation of the actual chemical species present in solution. It was known that calcium chloride and sodium phosphate were soluble in water and formed solutions because of the (aq) symbol written after them on the reactant side. The same is true of sodium chloride on the product side. It is also known that calcium phosphate is insoluble in water and thus precipitates because of the symbol (s) written after it on the product side.

Still a third way of writing this reaction equation is called the "**net ionic**" equation. In this form, all chemical species, which exist identically on both sides of the reaction equation, cancel and are deleted from the equation. In this case, it means $6Na^+(aq)$ and $6Cl^-(aq)$. These ions are given the name "**spectator ions**." They are present in solution but are not necessary for the reaction to occur.

Hence, the following equation is left: $3Ca^{2+}(aq) + 2PO_4^{3-}(aq) \rightarrow Ca_3(PO_4)_2(s)$.

It is noteworthy that while spectator ions do not participate in the net reaction, their presence contributes to the overall ionic strength of the solution. If large enough, their presence can increase the solubility of an insoluble salt. The effect is referred to as the "uncommon ion" or "salt" effect.

### 2.4.1 AN ILLUSTRATIVE EXAMPLE OF A DECOMPOSITION REACTION—AN EXPLOSION

Nitroglycerin, $C_3H_5N_3O_9$, is a well-known, powerful, liquid explosive. The explosion of nitroglycerin falls under the category of a decomposition reaction. Examining its products, the number of molecules, and the physical states they are in, as well as the amount of heat energy released, labeled $\Delta H$, explains its tremendous destructive power. The heat energy released is also known as the enthalpy of reaction, which will be discussed in more detail in Section 2.7. Consider the balanced reaction for this decomposition below:

$$4C_3H_5N_3O_9(l) \rightarrow 6N_2(g) + 12CO_2(g) + 10H_2O(g) + O_2(g) \quad \Delta H° = -5678\,kJ$$

(2.7)

Note the following three points in Equation 2.7:

1. A liquid reactant changes (upon ignition or shock) into all gaseous products, requiring greater volume.
2. Four moles of reactant decompose into 29 moles of products, requiring more volume.
3. A tremendous amount of heat energy—5678 kJ—is released (indicated by the negative sign of $\Delta H$), heating the already gaseous products, and forcing them to expand, requiring greater volume, i.e., increased temperature causes increased pressure, at constant volume.

This problem underscores the importance of understanding several individual chemical principles at work simultaneously.

## 2.5 LIMITING REAGENT

In many reactions, the reactants are not present in stoichiometric ratios. One of the reactants, called the **limiting reagent**, is present in short supply. This reactant determines the outcome of the reaction, that is, the maximum amount of any product of interest that can be generated. The other reactant or reactants are thus present **in excess**. In any reaction where two or more reactants are present and their respective amounts are given, the limiting reagent must be identified before the maximum amount of any product can be calculated.

**Example 2.2:** (A & B)Stoichiometry Calculation of a Limiting Reagent Reaction

In an experiment, 50 g of hydrogen gas and 50 g of oxygen gas are ignited to form water. Calculate:

A. The mass of the water formed.
B. The mass of any hydrogen or oxygen that is left over or unreacted.

*Solution*

First, write the balanced chemical reaction as follows:

$$2H_2 + O_2 \rightarrow 2H_2O$$

Since the amounts of both reactants are given, one of the reactants may be the limiting reagent. To find which reactant is the limiting reagent, convert each mass given in grams into moles, and compare. Thus:

$$50 \text{ g H}_2 \ \times \ \frac{1.0 \text{ mol H}_2}{2.016 \text{ g H}_2} = 24.8 \text{ mol H}_2$$

$$50 \text{ g O}_2 \ \times \ \frac{1.0 \text{ mol O}_2}{32 \text{ g O}_2} = 1.56 \text{ mol O}_2$$

From the balanced reaction, it is clear that 2 mol of $H_2$ requires 1 mol of $O_2$. Thus, 24.8 mol of $H_2$ would require 12.4 mol of $O_2$. However, only 1.56 mol of $O_2$ is actually present or available for reaction. Thus, $O_2$ is in short supply and is the limiting reagent. Also, $H_2$ is present in excess and will be left over after the reaction.

A. To find the mass of water produced, again apply the factor-label method, using the molar amount of $O_2$ present as the starting point.

$$\text{mol O}_2 \ \times \ \frac{2 \text{ mol H}_2\text{O}}{1 \text{ mol O}_2} \ \times \ \frac{18.0 \text{ g H}_2\text{O}}{1.0 \text{ mol H}_2\text{O}} = 56.2 \text{ g H}_2\text{O}$$

B. To find the mass of $H_2$ left over, subtract the amount of $H_2$ reacted with $O_2$ from the total mass of $H_2$ initially present and available in moles, since $O_2$ is the limiting reagent.
    24.8 mol $H_2$ initially present
    3.12 mol $H_2$ reacted (equivalent to $2 \times 1.56$ mol $O_2$)

$$21.68 \text{ mol H}_2 \text{ left over or unreacted} \ \times \ \frac{2.016 \text{ g H}_2}{1.0 \text{ mol H}_2} = 43.7 \text{ g H}_2$$

### 2.5.1  PERCENT YIELD

The result obtained in Example 2.1 (A) is the theoretically predicted maximum amount of water. In reality, the amount of water actually collected may, for at least two reasons, be somewhat less. First, some reactions may not go to completion but instead may reach an equilibrium condition. Second, other reactions may have more than one pathway and may produce secondary or tertiary products that compete with the main route. In a sense, the percent yield of a reaction is a measure of efficiency.

In either case, a **percent yield** may be computed as follows:

$$\%\text{Yield} = \frac{\text{actual yield}(\text{g})}{\text{theoretical yield}(\text{g})} \times 100$$

**Example 2.3:** Calculation of Percent Yield of a Reaction

Suppose that, in Example 2.2, the actual mass of water collected or measured is 48.0 g. Compute the percent yield of the reaction.

*Solution*

$$\%\,\text{Yield} = \frac{48.0\,\text{g}}{56.2\,\text{g}} \times 100 = 85.4\%$$

## 2.6   CONSECUTIVE AND SIMULTANEOUS REACTIONS

Reactions that are carried out one after another in sequence to yield a final product are called **consecutive reactions**. In **simultaneous reactions**, two or more reactants react independently of each other in separate reactions at the same time. Said differently, a reaction may have more than one pathway to produce a spectrum of products.

An example of a consecutive reaction involves the purification of titanium dioxide, $TiO_2$, a substance used in the manufacturing of such products as white pigment in paints, thin films on coated textiles, printing inks, sunscreens in cosmetics, soap, toothpaste, and photocatalyst agents. To free $TiO_2$ of unwanted colored impurities, it must first be converted into $TiCl_4$ and then reconverted into $TiO_2$.

$$2TiO_2\,(s)\ \left(\text{impure}\right) + 3C(s) + 4Cl_2\,(g) \rightarrow 2TiCl_4\,(g) + CO_2\,(g) + 2CO(g)$$

$$TiCl_4\,(g) + O_2\,(g) \rightarrow TiO_2\,(s)\left(\text{pure}\right) + 2Cl_2\,(g)$$

Note that the $TiCl_4$ product in the first reaction becomes a reactant in the second. This is the connecting link. For example, one could ask how many grams of carbon are required to produce 1.0 kg of pure $TiO_2$. First, it must be realized that 1 mole of $TiO_2$ (pure) requires 1 mole of $TiCl_4$. Then 2 moles of $TiCl_4$ requires 3 moles of carbon. In this way, using the factor-label method, a calculation with conversion factors can be set up, connecting $TiO_2$(s) (pure) with C(s).

## 2.7   ENERGY CHANGE FOR EXOTHERMIC VS. ENDOTHERMIC REACTIONS

Exothermic reactions are reactions that produce or generate heat to the environment or surroundings. By convention, their $\Delta H$ values are always negative. Endothermic reactions, on the other hand, absorb heat from the environment or surroundings. Their $\Delta H$ values are always positive. Units of heat energy are usually in either kilo-Joules per mole (kJ/mole) or kilo-calories per mole (kcal/mol). The $\Delta H$ value of a reaction is the difference in energy between the energy of the products and the energy of the reactants as shown in the energy reaction diagram (Figure 2.1).

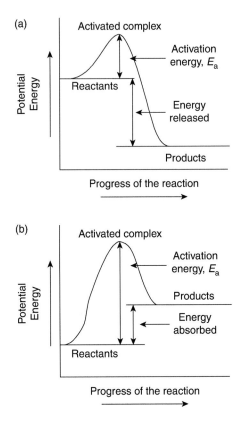

**FIGURE 2.1** Change in potential energy for (a) exothermic reaction and (b) endothermic reaction.

# 3 Acids, Bases, and Salts

## 3.1 ACIDS AND BASES

Some substances have special properties and are so frequently encountered that they are grouped in special classes and deserve separate treatment. This is the case with acids, bases, and salts.

According to the **Arrhenius model**, an **acid** is any substance that, when dissolved in water, produces or causes to produce hydrogen ions, $H^+$. A **base** is any substance that, when dissolved in water, produces or causes to produce hydroxide ions, $OH^-$. The **Bronsted–Lowry model** stresses the fact that acids are *proton donors* and bases are *proton acceptors*, as well as the notion of *conjugate* (meaning "related") acids and bases. The **Lewis model** states that acids are substances that are electron-pair acceptors and bases are substances that are electron-pair donors. This concept broadens the meaning of acids and bases considerably to include many salts that would otherwise be excluded. Realistically, since a hydrogen ion is a bare proton, it is too reactive to exist as a stable species by itself. Hence, the real acid species is the hydronium ion, $H_3O^+$, a protonated water molecule, that is, $[H^+(H_2O)]$. Acids are powerful dehydrating agents.

When an acid and a base are added together, a neutralization (a double-replacement reaction) ensues, with water and a salt as the neutralization products. Quantitatively, one mole of acid, $H_3O^+$, exactly neutralizes one mole of base, $OH^-$. Thus, this is true for monoprotic acids and monobasic bases. Monoprotic acids contain one $H^+$ ion, such as $HNO_3$, while diprotic acids contain two $H^+$ ions, such as $H_2SO_4$, and thus require two $OH^-$ ions to neutralize them. In general, polyprotic acids contain more than one ionizable $H^+$ ion; examples include phosphoric acid, $H_3PO_4$, with three $H^+$ ions, oxalic acid, $C_2O_4H_2$, with two $H^+$ ions, as well as the aforementioned sulfuric acid.

Acids and bases may be strong or weak. This property is determined by molecular structure and is different from concentration. A strong acid is one that dissociates and ionizes completely or 100% into its hydrogen or hydronium ions and corresponding anions, known as its conjugate bases. Similarly, a strong base is one that dissociates and ionizes completely into hydroxide ions and its corresponding cations, known as its conjugate acids. Weak acids and bases, in contrast, dissociate and ionize much less than 100%, conventionally less than 10%, and typically less than 5%, into their respective ions. Note that these substances are referred to as weak electrolytes, because their concentrations of individual ions are small, while strong acids and bases are strong electrolytes.

There are four common strong acids. All others can be assumed, by exclusion, to be weak.

DOI: 10.1201/9781003092759-3

- **Hydrochloric acid, HCl** [HBr and HI are also strong but uncommon; HF is weak.]
- **Nitric acid, HNO$_3$** [This is a *monoprotic acid* since it has only one hydrogen ion.]
- **Sulfuric acid, H$_2$SO$_4$** [This is a *diprotic acid* since it has two hydrogen ions.]
- **Perchloric acid, HClO$_4$**

Examples of common weak acids include acetic acid (in vinegar), boric acid (in the medicine cabinet), hydrofluoric acid (used as an agent to etch glass), ascorbic acid (vitamin C), carbonic acid, and phosphoric acid (the latter two in many carbonated beverages).

Examples of strong bases are sodium hydroxide (NaOH) and potassium hydroxide (KOH). Examples of weak bases are ammonia, NH$_3$, which, when dissolved in water, becomes ammonium hydroxide and may be written as NH$_4$OH, and organic amines, such as methyl amine, CH$_3$NH$_2$.

(For more a detailed discussion and calculations of weak acids and bases, please see the appropriate section of Chapter 4.)

When dissolved in water, **oxides of metals** (e.g., CaO, K$_2$O, MgO) undergo hydrolysis and are generally basic, while **oxides of nonmetals** (e.g., CO$_2$, SO$_2$, SO$_3$) also undergo hydrolysis and are generally acidic. Hydrolysis simply means reaction with water.

## 3.2   CONCENTRATION UNITS AND THE pH SCALE

Two units are routinely used to express concentrations of acids and bases:

- Molarity, M
- pH scale

Molarity is defined as the number of moles of H$^+$ ion for acids or of OH$^-$ ion for bases per liter of solution. Concentration is symbolized by the use of brackets; that is, [ ] means moles per liter of the bracketed quantity.

The **pH**, representing the power of hydrogen ions, can be calculated from molarity. pH is useful for solutions whose acidic concentrations are less than 1.0 M, for example, $3.5 \times 10^{-3}$M. This is the case for many solutions encountered in consumer products, biochemistry and biological applications as well as biomedical and environmental engineering.

$$pH = -\log\left[H^+\right] \tag{3.1}$$

Equation 3.1 may also be written as follows:

$$\left[H^+\right] = 10^{-pH} \tag{3.2}$$

While not an absolute restriction, the pH scale conventionally runs from 0 to 14, because of the autoionization of water. Recall that $K_w$ for water is $1.0 \times 10^{-14}$ at 25°C, since $K_w = [H^+][OH^-] = 1.0 \times 10^{-14}$.

The pOH may be analogously computed for basic solutions:

$$pOH = -\log\left[OH^-\right] \qquad (3.3)$$

In addition, since

$$pH + pOH = 14.0 \qquad (3.4)$$

it is always possible to find pOH from pH, or vice versa.

For the more commonly used pH, the **acid range** of solutions is **0–7**, while the **basic** or **alkaline** range is **7–14**. A pH of 7.0 represents a neutral solution such as distilled water in which the hydrogen ion concentration and the hydroxide ion concentration are both equal to $1.0 \times 10^{-7}$ M. The pH values of common substances are shown in Figure 3.1.

Properties of acids include the following:

- They taste sour.
- They neutralize bases to produce water and a salt.
- They react with many metals to yield $H_2$ gas.
- They are strong dehydrating agents.

Properties of bases include the following:

- They taste bitter.
- They feel slippery.
- They react with fats and oil.
- They neutralize acids to produce water and a salt.

| Concentration of Hydrogen ions compared to distilled water | | | Examples of solutions and their respective pH |
|---|---|---|---|
| 1/10,000,000 | 14 | Liquid drain cleaner, Caustic soda | |
| 1/1,000,000 | 13 | bleaches, oven cleaner | |
| 1/100,000 | 12 | Soapy water | |
| 1/10,000 | 11 | Household Ammonia (11.9) | |
| 1/1,000 | 10 | Milk of magnesium (10.5) | |
| 1/100 | 9 | Toothpaste (9.9) | |
| 1/10 | 8 | Baking soda (8.4), Seawater, Eggs | |
| 0 | 7 | "Pure" water (7) | |
| 10 | 6 | Urine (6) Milk (6.6) | |
| 100 | 5 | Acid rain (5.6) Black coffee (5) | |
| 1,000 | 4 | Tomato juice (4.1) | |
| 10,000 | 3 | Grapefruit & Orange juice, Soft drink | |
| 100,000 | 2 | Lemon juice (2.3) Vinegar (2.9) | |
| 1,000,000 | 1 | Hydrochloric acid secreted from the stomach lining (1) | |
| 10,000,000 | 0 | Battery Acid | |

**FIGURE 3.1**   The pH of some common substances.

## 3.3   CALCULATIONS FOR STRONG ACIDS AND BASES

Examples 3.1–3.8 represent routinely encountered problems and calculations for strong acids and bases. In general, 1 mole of acid ($H^+$ ion) neutralizes 1 mole of base ($OH^-$ ion) to produce 1 mole of water ($H_2O$).

**Example 3.1:** (A & B) Understanding Molarity of a Strong Acid

An aqueous solution of hydrochloric acid is 0.034 M. Find:

    A. $[H^+]$
    B. $[Cl^-]$

### Solution

Since HCl is a strong acid, it dissociates and ionizes completely (i.e., 100%) into its component ions. Thus,

    A. $[H^+] = 0.034\,M$
    B. $[Cl^-] = 0.034\,M$It is safe to assume that [HCl] in terms of molecules is 0.

**Example 3.2:** (A & B) Molarity and pH Calculations of a Strong Acid

For a 0.035 M $HNO_3$ solution, find:

    A. $[H^+]$
    B. pH

### Solution

    A. $HNO_3$ is a strong acid. Thus, a 0.035 M solution of $HNO_3$ yields 0.035 M $H^+$ ions (and 0.035 M $NO_3^-$ ions), since there is a complete dissociation and ionization.
       So, $[H^+] = 0.035\,M$
    B. $pH = -\log[H^+] = -\log\,(0.035\ M) = 1.46$

**Example 3.3:** Molarity Calculation for a Base

Find $[OH^-]$ for the solution in Example 3.2.

### Solution

Since this is an aqueous solution,

$$\left[H^+\right]\left[OH^-\right] = K_w = 1.0 \times 10^{-14} \tag{3.5}$$

Thus,

$$\left[OH^-\right] = \frac{K_w}{\left[H^+\right]} = \frac{1.0 \times 10^{-14}}{0.035\,M} = 2.86 \times 10^{-13}\ M$$

This shows a very small OH⁻ ion concentration, but still not zero, and illustrates the constant, reciprocal nature of $H^+$ ions and $OH^-$ ions in an aqueous solution.

### Example 3.4: Determining the pH of Lemon Juice

A sample of lemon juice has a pH of 2.4. Find its [H⁺].

*Solution*

Recall that [H⁺] = $10^{-pH}$. Thus:

$$\left[H^+\right] = 10^{-2.4} = 4.0 \times 10^{-3} \text{ M}$$

Note that this is consistent with the range of predicted values for acidic substances in Figure 3.1.

### Example 3.5: (A & B) Molarity and pH Calculations of an Alkaline Solution

Find, for a 0.014 M solution of slaked lime, $Ca(OH)_2$:

    A. [OH⁻]
    B. pH

*Solution*

    A. The subscript 2 next to (OH⁻) indicates that 1 mol of $Ca(OH)_2$ produces
       2 mol of OH⁻ ions (along with 1 mol of $Ca^{2+}$ ions).
       It is assumed that $Ca(OH)_2$ is a strong base.
       Thus, [OH⁻] = 2(0.014 M) = 0.028 M
    B. Furthermore, pOH = −log [OH⁻] = −log(0.028 M) = 1.55. Then,
       pH = 14.0 − pOH = 14.0 − 1.55 = 12.45.

### Example 3.6: Finding the pH of an Acid-Base Reaction

A laboratory technician mixes 400 mL of a 0.125 M NaOH solution with 600 mL of a 0.100 M HCl solution. Find the pH of the resulting solution.

*Solution*

First compute the number of moles of acid and base to determine whether they are equal or, if not, which one is present in excess.
    Moles of H⁺ = (volume HCl)(molarity HCl) = 0.060 mol
    Moles of OH⁻ = (volume NaOH)(molarity NaOH) = 0.050 mol
    Since 0.06 mol of H⁺ > 0.05 mol of OH⁻, the final solution will be acidic.
    Furthermore, since 1 mol of acid reacts exactly with 1 mol of base, subtract moles of base from moles of acid to find the net mols of acid: 0.060 mol − 0.050 mol = 0.010 mol of H⁺ left over or unreacted and present after mixing. This amount is in a total, combined volume of 1000 mL or 1.00 L. Thus,

$$\left[H^+\right] = \frac{0.01\,\text{mol H}^+}{1.0\,\text{L}} = 0.01\,M$$

$$pH = 2.0$$

**Example 3.7:** Finding the Concentration (M) of an Unknown Base by Titration

A laboratory technician wishes to find the concentration of an unknown base. He performs a titration in which 42.50 mL of 0.150 M HCl exactly neutralizes 25.00 mL of the base. Determine the concentration of the unknown base.

*Solution*

Since moles of acid (A) equal moles of base (B) at the *endpoint* (point of neutralization), use the relationship

$$V_A M_A = V_B M_B \tag{3.6}$$

Solve for $M_B$:

$$M_B = \frac{V_A M_A}{V_B} = \frac{\left(42.50\,\text{mL}\right)\left(0.150\,M\right)}{25.00\,\text{mL}} = 0.255\,M$$

**Example 3.8: Dilution of a Solution**

A laboratory technician is asked to prepare 500 mL of a 0.750 M solution of HCl. The stock solution of HCl that she has is labeled 6.0 M. How much (what volume) should she take from the stock solution bottle?

*Solution*

The total number of moles of HCl ultimately desired in solution is:
    Moles of HCl = $V_{HCl}M_{HCl}$ = (0.500 L)(0.750 mol/L) = 0.0375 mol
    This is the amount that must come from the 6.0 M solution. Since this amount is simply being redistributed from a concentrated solution ($M_1$ and $V_1$) to a more dilute one ($M_2$ and $V_2$), the number of moles of HCl in solution must remain constant. Thus, the number of moles of HCl *before* (1) must equal the number of moles of HCl *after* (2). So the calculation goes as follows:

$$M_1 V_1 = M_2 V_2 \tag{3.7}$$

Solve for $V_1$, substituting the values:

$$V_1 = \frac{\left(0.750\,M\right)\left(0.500\,L\right)}{6.0\,M} = 0.0625\,L = 62.5\,\text{mL of stock solution}$$

This same strategy can be applied to *any* dilution problem.

## 3.4   SALTS AND COLLIGATIVE PROPERTIES

In general, salts are substances composed of a cation and an anion. They are usually encountered as products in acid–base neutralization reactions. They may be strong or weak electrolytes; they may vary in solubility with water. Those that are very soluble in water, such as sodium chloride, are said to be completely soluble, even though they do have an upper limit of solubility. Those that are only partially soluble in water are said to be insoluble, even though they do have a very small, limited solubility in water, e.g., barium sulfate. When these salts are dissolved in water, they form solutions with interesting properties, usually referred to as colligative properties. A solution is composed of a solute, such as sodium chloride, dissolved in a solvent, such as water. A solute must be completely dissolved in a sufficient volume of solvent with no identifiable, visible solute for it to qualify as a solution.

There are four important colligative properties of solutions:

- Boiling-point elevation
- Freezing-point depression
- Vapor-pressure lowering
- Osmotic pressure

The term "colligative" means having to do with the *collection* or *number* of particles and/or ions dissolved in solution. It is an oversimplification, since size and polarity also play a role, albeit a less important one, in the chemistry of solutions. Colligative properties depend on the concentration of the solute (in units of molality) and its "*i*" factor (**van't Hoff factor**), i.e., the total number of dissolved particles. The "*i*" factor depends on the specific chemical formula of the solute as well as the type of solute, as discussed below under "electrolytes."

Before describing each of the four colligative properties or effects in detail, it is useful to review the concept of a solution. Recall that a solution is a physical state wherein a solute has been dissolved completely and uniformly in a solvent, forming a solution of uniform composition. Such a solution is said to be *homogeneous*. There is only one identifiable phase. Unless specified otherwise, the solvent is assumed to be water, whose normal boiling point, freezing point, and density are well known. In the equations for each of the four properties, the concentration of the solution appears, although concentration units may vary. It is critically important, when doing calculations with solutions, to distinguish among units of mass, volume, and density for solute, solvent, and solution, respectively.

It is useful to review the different concentration units quantify concentrations of aqueous solutions originally presented in Figure 1.2 from Chapter 1. Five common definitions and concentration units are presented here again in Figure 3.2. The reason for the great variety is that different applications and chemical formulas require different units to keep these formulas mathematically manageable. The term "concentration" always implies the existence of a **solute**, present as a gas, liquid, or solid, dissolved in a suitable **solvent** to form a **solution**. Unless otherwise specified,

> **Molarity, M** = moles of solute/liter of solution.
> **Molality, m** = moles of a solute/kilogram of solvent.
> **Mole Fraction, X** = number of moles of component *i*/total number of moles of all species.
> **Percent by Mass/Mass** = mass of solute in grams/mass of solution in grams.
> **Percent by Volume/Volume** = volume of solute in milliliters/total volume of solution in milliliters.

**FIGURE 3.2**    Common concentration units.

the solvent for liquid solutions is understood to be water—hence the designation "aqueous solution," abbreviated as **(aq)**.

For solutions, there are three types of solutes or "electrolytes":

- Strong electrolytes
- Nonelectrolytes
- Weak electrolytes

Strong electrolytes are generally ionic compounds, such as salts, and strong acids and bases, that dissociate and ionize completely, that is, 100%, in water. The solutions they form with water become strong conductors of electric currents. Examples include NaCl, $KNO_3$, $CaCl_2$, HCl, and NaOH. Thus, 1 mole of NaCl produces 2 moles of ions or particles; 1 mole of $CaCl_2$ produces 3 moles of ions or particles (1 mole of $Ca^{2+}$ ions and 2 moles of $Cl^-$ ions. The **van't Hoff factor** is the total number of ions or particles per formula unit or mole: 2.0 for NaCl, 3.0 for $CaCl_2$, and so on. Experimentally measured van't Hoff factors are generally less than theoretically predicted ones for a given solute.

Nonelectrolytes are generally alcohols and sugars, such as ethanol ($C_2H_5OH$) and glucose ($C_6H_{12}O_6$), or other organic substances that dissolve in water but do not ionize. One mole of a nonelectrolyte always produces 1 mole of particles. Thus, the van't Hoff factor is 1.0.

A weak electrolyte, such as acetic acid, is a substance that is only partially dissociated and ionized, typically 10% or less, leaving 90% or more undissociated and thereby behaving much like a nonelectrolyte. Its van't Hoff factor is greater than 1 but less than the total number of ions available for dissociation.

## 3.5   BOILING-POINT ELEVATION

When a nonvolatile solute is dissolved in water or another suitable solvent, the normal boiling point of the solvent is always raised or elevated. The amount of elevation in temperature, $\Delta T$, is given by the equation

$$\Delta T = iK_b m \tag{3.8}$$

where
    $i$ = the van't Hoff factor, discussed above,
    $m$ = the molality of the solution

## TABLE 3.1
## Cryoscopic and Ebullioscopic Constants for Common Solvents

| Substance | Freezing Point (°C) | Cryoscopic Constant (K·kg/mol) | Boiling Point (°C) | Ebullioscopic Constant (K·kg/mol) |
|---|---|---|---|---|
| Acetic acid | 16.6 | 3.90 | 118.1 | 3.07 |
| Benzene | 5.5 | 5.12 | 80.1 | 2.53 |
| Camphor | 179.8 | 39.7 | 204.0 | 5.95 |
| Cyclohexane | 6.4 | 20.2 | 80.74 | 2.79 |
| Diethyl ether | −114.3 | 1.79 | 34.5 | 2.16 |
| Ethanol | −114.6 | 1.99 | 78.4 | 1.19 |
| Water | 0.0 | 1.86 | 100.0 | 0.512 |

$K_b$ = the ebullioscopic or boiling-point elevation constant, characteristic of water (or other solvent) in °C · kg/mol. See Table 3.1, which lists $K_b$ and $K_f$ values for common solvents.

**Example 3.9:** Boiling Point Elevation of a Salt Solution

What is the boiling point of a solution made by dissolving 70 g of NaCl in 300 g of water? The $K_b$ for water is 0.512°C · kg/mol.

*Solution*
Use Equation 3.8: $\Delta T = iK_b m$ and note that $i = 2$ for NaCl. Then, molality,

$$m = \frac{\text{mol NaCl}}{\text{kg H}_2\text{O}}$$

$$= \frac{70 \text{ gNaCl} / 58.5\text{g} / \text{mol}}{0.300 \text{ kg H}_2\text{O}} = 3.99$$

$$\Delta T = (2)(0.512)\left(3.99 \text{ mol} / \text{kg}\right) = 4.08°C$$

$$T_{\text{new}} = T_{\text{norm}} + \Delta T = 100°C + 4.08°C = 104.1°C$$

## 3.6 FREEZING-POINT DEPRESSION

When a nonvolatile solute is dissolved in water or any other solvent, the normal freezing point of the solvent is always lowered or depressed. The amount of depression in freezing point, $\Delta T$, is given by the equation

$$\Delta T = iK_f m \tag{3.9}$$

where $i$ and $m$ have the same definitions as in Equation 3.8, and $K_f$ is the cryoscopic or freezing-point depression constant in °C kg/mol. See Table 3.1 for a list of $K_f$ values for common solvents.

**Example 3.10:** Determining the Molar Mass by Freezing Point Depression

The freezing-point depression equation is often used to find the molar mass of an unknown solute. Find the molar mass (also known as the molecular weight) of a substance when 1.14 g of it is dissolved in 100.0 g of liquid camphor, whose freezing point is lowered by 2.48°C. $K_f$ for camphor is 39.7°C kg/mol as given in Table 3.1.

*Solution*

Use Equation 3.9: $\Delta T = iK_f m$. Assume that the unknown substance is an organic nonelectrolyte, since camphor is also an organic nonelectrolyte, and the principle of like dissolves like applies. Thus, $i = 1$. Then,

$$\Delta T = 2.48°C \text{ and } K_f = 39.7°C \text{ kg/mol}$$

First, solve for the molality, m:

$$m = \frac{\Delta T}{iK_f} = \frac{2.48°C}{(1)(37.9)} = 0.0625\,\text{mol}\,/\,\text{kg}$$

But

$$m = \frac{\text{mol solute}}{\text{kg solvent}} = \frac{\text{grams solute/molar mass solute}}{\text{kg solvent}}$$

Solve for molar mass, *MM*, of the solute:

$$MM = \frac{\text{g solute}}{(\text{kg solvent})(m)} = \frac{1.14\,\text{g}}{(0.100\,\text{kg})(0.0625\,\text{mol/kg})}$$

$$= 182\,\text{g/mol}$$

This number can then be compared with a hopefully short list of potential unknown substances, whose molar mass, as determined by the chemical formula, best matches this number. Note that for this type of analysis, solvents are chosen, which have a large $K_f$ value to help minimize the relative error from experiment.

## 3.7　HENRY'S LAW

**Henry's Law**, while not, strictly speaking, considered a colligative property, describes the relationship of the solubility of a gas in a liquid, usually water. The law states that the solubility, *S*, of a gas in a liquid is directly proportional to the partial pressure, *P*, of the gas above the liquid; it can be written as follows:

$$S = k_H P \qquad (3.10)$$

where $k_H$ is called Henry's Constant. Henry's Constant depends on the gas, the solvent, and the temperature. A table of values for $k_H$ may be found in the Handbook of

**TABLE 3.2**
**Henry's Law Constants $k_H$ for Aqueous Solutions at 25°C**

| Gas | Constant $k_H$ (Pa/(mol/liter)) | Constant $k_H$ (atm/(mol/liter)) |
|-----|----------------------------------|-----------------------------------|
| He | $282.7 \times 10^{+6}$ | 2865 |
| $O_2$ | $74.68 \times 10^{+6}$ | 756.7 |
| $N_2$ | $155 \times 10^{+6}$ | 1600 |
| $H_2$ | $121.2 \times 10^{+6}$ | 1228 |
| $CO_2$ | $2.937 \times 10^{+6}$ | 29.76 |
| $NH_3$ | $5.69 \times 10^{+6}$ | 56.9 |

Chemistry and Physics as well as in many introductory textbooks on chemistry. A brief table of $k_H$ values for common gases is given in Table 3.2. The law implies that, at constant temperature, doubling the partial pressure of a gas doubles its solubility.

Henry's Law helps explain such phenomena as the bends experienced by deep-sea divers who surface too quickly as well as the carbonation of beverages in tightly capped bottles. The reciprocal of Henry's Constant may be used as a measure of the volatility of organic substances (e.g., hexane or octane) dissolved in water, albeit with a limited solubility in water.

Another interpretation or extension of $k_H$ is that it is a kind of partition coefficient. It expresses the partitioning of a substance between two different phases. It is the ratio of the concentration of a gas in the vapor phase (in air) to that in the solvent phase. More generally, a **partition coefficient**, $K_p$, is defined as the ratio of concentrations of a compound between two immiscible solvents, e.g., the concentration (or solubility) of caffeine in chloroform and its concentration (or solubility) in water. The $K_p$ for this particular example in fact is 9.1 at 25°C. This simply means that caffeine is about nine times as soluble in chloroform as it is in water. Note that partition coefficients are unitless. They are often expressed in terms of natural logarithms, e.g., ln $K_p$. Clearly, they are of great value in calculating and predicting the concentration of an organic compound, e.g., benzene, a pesticide, or other chlorinated hydrocarbon, distributed between an organic solvent and water, keeping in mind that the organic solvent may be a single solvent like hexane or a mixture like gasoline.

More generally, partition coefficients are of great value in predicting or at least estimating the *effective* molar concentration of an organic compound (e.g., benzene or octane) in physical contact with an inorganic solvent (e.g., water) as in heterogeneous, air–soil or water–soil systems. If gasoline were to leak or spill into a static aquatic system, e.g., a holding pond or aquifer, despite its limited solubility in water, a portion of its primary components (e.g., benzene or octane) would dissolve in the water. For further discussion and examples of calculations involving partition coefficients, the interested reader should consult *Chemical Property Estimation* by Edward Baum, CRC Press, 1998.

## 3.8 RAOULT'S LAW FOR VAPOR-PRESSURE LOWERING

**Raoult's Law** states that when a nonvolatile solute is added to water or another solvent, forming a solution, the vapor pressure of the solvent is depressed. In other words, the vapor pressure of the solution is lower than the vapor pressure of the pure solvent. The vapor pressure of the solution is directly proportional to the mole fraction of the solvent multiplied by the vapor pressure of the pure solvent:

$$P_{soln} = X_{solvent}\, P^{\circ}_{solvent} \tag{3.11}$$

where $P_{soln}$ = the vapor pressure of the solution,

$$X_{solvent} = \text{the mole fraction of the solvent} = \frac{\text{mol solvent}}{\text{mol solvent} + \text{mol solute}}$$

$P^{\circ}_{solvent}$ = the vapor pressure of the pure solvent.

Vapor pressures are a measure of volatility. They may be expressed in any suitable units, such as atmospheres, millimeters of mercury, torr, and pounds per square inch.

Note that Raoult's Law applies to all ideal solutions. Ideal solutions mean systems where solute and solvent molecules have similar molecular structures or geometries and thus experience similar intermolecular forces (as in a benzene–toluene solution, a system of two structurally similar organic molecules, for example), *OR* systems that are dilute solutions (where $X_{solvent} > 0.85$). For a binary system with only one solute and one solvent, Raoult's Law may alternatively be expressed as follows:

$$P_{soln} - P_{solvent} = \Delta P = X_{solute} P_{solvent}^{\circ} \tag{3.12}$$

Since

$$X_{solvent} + X_{solute} = 1 \tag{3.13}$$

**Example 3.11:** Finding Vapor Pressure of a Sugar Solution Using Raoult's Law

Calculate the vapor pressure at 100°C of a solution prepared by dissolving 10 g of sucrose, $C_{12}H_{22}O_{11}$, in 100 g of water.

### Solution

This is a real solution that is dilute, so Raoult's Law should apply.

$$P_{soln} = X_{solvent} P_{solvent}^{\circ} = X_{H_2O} P_{H_2O}^{\circ} \tag{3.14}$$

Where $P_{H_2O}^{\circ} = 760$ torr since the normal boiling point of water is 100°C, and its vapor pressure at that temperature must be equal to atmospheric pressure. On the other hand, the vapor pressures of pure solvents can always be looked up in tables.

**TABLE 3.3**
**Vapor Pressure of Water as a Function of Temperature**

| Temperature (°C) | Vapor Pressure (Torr) | Temperature (°C) | Vapor Pressure (Torr) |
|---|---|---|---|
| 0 | 4.6 | 27 | 26.7 |
| 5 | 6.5 | 28 | 28.3 |
| 10 | 9.2 | 29 | 30.0 |
| 15 | 12.8 | 30 | 31.8 |
| 20 | 17.5 | 31 | 33.7 |
| 21 | 18.7 | 32 | 35.7 |
| 22 | 19.8 | 33 | 37.7 |
| 23 | 21.1 | 34 | 39.9 |
| 24 | 22.4 | 35 | 42.2 |
| 25 | 23.8 | — | |
| 26 | 25.2 | 100 | 760 |

Table 3.3 shows the values of the vapor pressures of some common solvents at room temperature, 25°C.

To find $X_{H_2O}$, compute the number of moles of sucrose *and* the number of moles of water.

$$\text{moles of sucrose} = 10\,\text{g}\,C_{12}H_{22}O_{11} \times \frac{1.0\,\text{mol}\,C_{12}H_{22}O_{11}}{342.3\,\text{g}\,C_{12}H_{22}O_{11}} = 0.0292\,\text{mol}$$

$$\text{moles of water} = 100\,\text{g}\,H_2O \times \frac{1.0\,\text{mol}\,H_2O}{18.0\,\text{g}\,H_2O} = 5.55\,\text{mol}$$

Thus,

$$X_{H_2O} = \frac{5.55\,\text{mol}}{5.55\,\text{mol} + 0.0292\,\text{mol}} = 0.9948$$

Finally,

$$P_{\text{soln}} = (0.9948)(760\,\text{torr}) = 756\,\text{torr}$$

Additionally,

$$\Delta P = 760\,\text{torr} - 756\,\text{torr} = 4.0\,\text{torr}$$

Vapor pressure is a function of temperature. The dependence of the vapor pressure of a pure solvent on temperature is given by the **Clausius–Clapeyron equation**:

$$\ln\frac{P_2}{P_1} = \frac{\Delta H_{\text{vap}}^{\,\circ}}{R}\left(\frac{1}{T_1} - \frac{1}{T_2}\right) \tag{3.15}$$

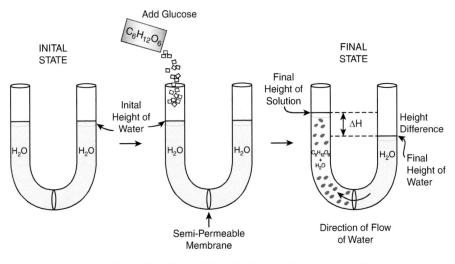

Equilibration is reached in Final State. At this point, the osmotic pressure developed
initally is equal to the hydrostatic pressure reached at the final state.

**FIGURE 3.3**   Three U-tubes showing the development of osmotic pressure.

where $P_1$ and $P_2$ are the vapor pressures at absolute temperatures $T_1$ and $T_2$,
respectively, and must be in Kelvins, and $\Delta H_{vap}°$ is the molar heat of vaporization of
the pure solvent at its *normal* or *standard* boiling point.

When the molar hear of vaporization or the boiling point of a substance is
unknown, a frequently useful, empirical approximation for many liquids, *if hydrogen
bonding is absent or unimportant*, is Trouton's Rule:

$$\frac{\Delta H_{vap}°}{T_b} = 85 \tag{3.16}$$

This ratio, expressed in joules per Kelvin, allows the calculation of either the standard
molar heat of vaporization, $\Delta H_{vap}°$, or the boiling point, $T_b$, when one or the other is
known (Figure 3.3).

## 3.9   OSMOTIC PRESSURE

Consider a U-tube in which a pure solvent such as water in the right arm is brought
into contact with an aqueous solution made with a nonvolatile solute such as a glucose
(or any other soluble substance) in the right arm, and the two arms are separated by
a semipermeable membrane (e.g., a selective membrane), permeable only to water,
i.e., solvent molecules. The two sides or arms of the U-tube are initially at the same
height. It is observed that a pressure develops across the membrane that drives solvent
molecules from the right arm into the left arm in an effort to dilute the solution and

try to equalize concentrations in both arms. This pressure is called **osmotic pressure**, $\Pi$, and the process is known as osmosis. It is given by the equation:

$$\Pi = iMRT \qquad (3.17)$$

where

$i$ = the van't Hoff factor, which is equal to 1.00 since glucose is a nonelectrolyte,
$M$ = the molarity of the solution,
$R$ = the universal gas constant, 0.0821 L $\cdot$ atm/K $\cdot$ mol,
$T$ = the absolute temperature in Kelvins.

Sometime later, the system in the U-tube reaches equilibrium, and a height differential, $\Delta H$, between the solution level on the left and pure solvent level on the right is observed. Because of this height differential, a hydrostatic pressure, $\Delta P$, develops and remains and is given by the equation

$$\Delta P = \rho g \Delta H \qquad (3.18)$$

where $\rho$ = the density of the glucose solution and $g$ = the acceleration constant due to gravity.

In fact, the final hydrostatic pressure is equal to the initial osmotic pressure. Note the basic difference between osmosis and diffusion. While both processes are driven by concentration gradients, a semipermeable membrane is present as a barrier. No barrier is present in the process of diffusion.

Osmosis is a commonly occurring biological process, regulating and maintaining the proper concentrations of electrolytes in the cells of the human body. Osmometry is often used to find the molar mass (MM) of an unknown substance dissolved in solution through its molarity, M, when the other quantities—mass of unknown substance, volume of solution, temperature, and osmotic pressure—can be measured experimentally.

A calculation of osmotic pressure is shown in the following example.

**Example 3.12:** Finding the Osmotic Pressure of a Sugar Solution

A very dilute sucrose solution of concentration 0.010 M in water is separated from pure water by an osmotic (i.e., semipermeable) membrane. Find the osmotic pressure that develops at 25°C (298 K).

### Solution

Use Equation 3.17, $\Pi = iMRT$ where $i = 1$ for sugar and all other nonelectrolytes.
So $\Pi = (1)(0.010\ M)(0.0821\ \text{L-atm/K-mol})(298\ K) = 0.24$ atm.
This result can also be converted into units of Torr, if desired, since 1.0 atm = 760 Torr.
In that case, the osmotic pressure would be 182 Torr.

## 3.10   COLLOIDS: DISPERSIONS AND SUSPENSIONS

In chemistry, mixtures may be classified as either homogeneous or heterogeneous. Homogenous mixtures are simply called solutions. Solutions have uniform or constant composition throughout the entire system. Only one phase is discernible. Solutions may be gaseous, like clean, dry air; liquid, like salt water; or solid, like many gemstones such as rubies or sapphires. Heterogeneous mixtures are called simply mixtures. There may be two or more phases discernible, each of which does NOT dissolve in the others. Motor oil and water, or sand and water, each constitute a liquid mixture. Plywood, wafer board, and particle board are examples of solid mixtures. In the real world, examples of mixtures far outnumber those of solutions.

However, there is an intermediate class of mixtures referred to as **colloids** or **colloidal systems, either dispersions or suspensions**. It should be noted that the four colligative properties equations previously discussed do not apply to colloidal systems. Each colloidal system has unique properties which must be determined experimentally.

A colloidal system forms when the solute particle, referred to as the dispersed phase, is relatively large compared with the solvent molecules, referred to as the dispersing medium. A colloidal system will usually form when the solute or particle to be dissolved (or dispersed) in a solvent, usually water, has one or more dimensions that fall in the range of 1 and 1000 nm or nm (1000 nm = 1 μm or 1 micron). Occasionally, the particle may be as large as 10 microns, e.g., macromolecules. When the dispersing medium is water, the colloidal system is often referred to as a **hydrocolloid**. Furthermore, hydrocolloids may be classified as either hydrophilic or hydrophobic. Hydrophilic colloids consist of particles attracted to water (sometimes referred to as reversible sols), while hydrophobic colloids consist of particles repelled by water (sometimes referred to as irreversible sols). The particles of the dispersed phase can take place in different phases depending on how much water (the dispersing medium) is available. Jell-O powder mixed with water is an example of a hydrocolloid. Medical dressings are creations of hydrocolloids as well.

A true colloidal dispersion is stable as it sits. Milk and blood are two examples. If a colloidal system requires periodic mixing or stirring, it is more properly referred to as a **colloidal suspension or sol**. Paints and cough medicines, which need to be shaken before use, are common examples.

An easy way to determine if a mixture is a colloidal dispersion is through the use of the Tyndall Effect. When a beam of light is shined through a true solution, the light passes cleanly through, i.e., in only one direction. However, when a light beam is shined through a colloidal dispersion, the substance in the dispersed phase scatters the light in all directions allowing it to be readily seen regardless of the position of the observer. An example of this is the shining of a flashlight in fog or milk, since each is an example of a colloid.

Examples of common colloidal systems organized in terms of dispersed phase and dispersed medium are given in Table 3.4. Note the use of refined terms by exact category: **emulsion, foam, aerosol, mist, gel, sol**.

**TABLE 3.4**
**Colloidal Systems**

| Dispersing Medium | Dispersed Phase | | |
|---|---|---|---|
| | **Gas** | **Liquid** | **Solid** |
| Gas (e.g., air) | Except for helium in xenon, none are known. | Liquid aerosols: fog; clouds; mist (sneeze, hair spray). | Smoke; welding fumes; fine dust/particulate matter (PM 2.5); ice clouds. |
| Liquid (e.g. water) | Foams: soap suds; whipped cream; shaving cream. | Emulsions: liquid crystal display; milk; mayonnaise; liquid biomolecular condensate; medical/surgical dressing. | Liquid suspensions (sols): paint; cough medicine; ink; solid biomolecular condensate. |
| Solid | Solid foam: aerogel; styrofoam. | Agar-Agar; gelatin (Jell-O); jelly; gel-like biomolecular condensate. | Solid suspensions (sols): toothpaste; cranberry glass. |

# 4 Chemical Equilibrium

## 4.1 BASIC CONCEPT

Reaction equations describe substances, called reactants, which when put together react and produce other different substances, called products. It may appear as if only products remain after the reaction is finished. In reality, many reactions do not go to completion, even if the reactants are present in stoichiometric ratios or amounts. Rather, they reach a condition known as equilibrium, denoted by double, reversible arrows in the reaction equation. **Equilibrium** means that there is a balance between the reactant side and the product side, or simply between the reactants and the products, and that the reaction is reversible. The chemistry of many air pollutants falls and many bodily functions under the heading of equilibrium reactions. The equilibrium condition is dynamic, not static, allowing microscopic changes in reactant and product concentrations to take place, such that no net change in reactant or product concentrations occurs, provided that no external stresses are applied. At any given time, all species in the reaction equation—reactants and products—are present at equilibrium in varying amounts. The relationship among these varying amounts can be described by a mathematical formula known as the equilibrium constant expression or simply the equilibrium expression. The equilibrium expression is set equal to an equilibrium constant symbolized by $K_c$.

An equilibrium reaction can be generally represented as follows:

$$a\mathrm{A} + b\mathrm{B} \rightleftharpoons g\mathrm{G} + h\mathrm{H} \tag{4.1}$$

The $K_c$ expression can then be expressed as follows:

$$K_c = \frac{[\mathrm{G}]^g [\mathrm{H}]^h}{[\mathrm{A}]^a [\mathrm{B}]^b} \tag{4.2}$$

where $a$, $b$, $g$, and $h$ represent the stoichiometric coefficients in the balanced reaction, and the brackets [ ] indicate molar concentrations. The simplest interpretation of $K_c$ is that it is a measure of the extent to which a reaction goes toward completion, i.e., a reaction where the product side is favored. The meaning of $K_c$ is discussed in more detail later in this section.

Four *points of caution* regarding the use of this expression are noteworthy:

- Only concentrations in units of molarity are permitted.
- Only species that have concentrations can appear. Thus, species in the solution state (aq) or in the gaseous state (g) are included, but species that are pure solids (s) or liquids (l) are not.
- $K_c$ is a function of temperature (check the van't Hoff Equation, Equation 4.5, later in this chapter).
- All concentrations substituted into this expression must be taken at equilibrium.

DOI: 10.1201/9781003092759-4

A reaction in which all species are present in the same physical state (e.g., all gases or all solutions) is called a **homogeneous reaction**, while a reaction in which species are present in more than one physical state (i.e., mixed physical states) is called a **heterogeneous reaction**. If all species in a reaction are present in the gaseous state, an alternative form of the equilibrium constant expression can be written as $K_p$:

$$K_p = \frac{P_G{}^g P_H{}^h}{P_A{}^a P_B{}^b} \qquad (4.3)$$

Here, $P$ represents the partial pressure of each gas present at equilibrium. Any pressure unit is acceptable as long as it is consistent with the others and with the units of $K_p$, but atmospheres and torr are common.

A formula to convert $K_c$ to $K_p$, and vice versa, is as follows:

$$K_p = K_c (RT)^{\Delta n} \qquad (4.4)$$

where
$T$ = the absolute temperature in Kelvins
$R$ = the gas constant to be used in units consistent with partial pressure
$\Delta n$ = the change in the number of moles of gas, i.e., the total moles of product *minus* the total moles of reactant in the balanced reaction

**Examples 4.1:** Expressing and Evaluating $K_c$ and $K_p$

Write the equilibrium constant expression $K_c$ for the following reaction at 25°C:

$$2N_2O_5(g) \rightleftharpoons 4NO_2(g) + O_2(g)$$

## Solution

Note that this is a homogeneous, gas-phase reaction. All three substances can have concentrations in units of molarity and should appear in the $K_c$ expression:

$$K_c = \frac{[NO_2]^4 [O_2]}{[N_2O_5]^2}$$

Given a value for $K_c$ and when asked to convert it to $K_p$, then, since all species are in the gaseous state, the equation would be the following:

$$K_p = K_c (RT)^{\Delta n}$$

$$= K_c (0.0821 \text{L.atm/K.mol} \times 298\text{K})^3$$

since $\Delta n = (4+1) - 2 = 3$

Note that although the temperature is given, it is not used in the calculation. It may be useful in other calculations, however.

**Examples 4.2:** Expressing and Evaluating $K_c$ and $K_p$

Sulfur dioxide, $SO_2(g)$, is a common air pollutant. For the reaction

$$2SO_2(g) + O_2(g) \rightleftharpoons 2SO_3(g)$$

At 827°C, the value for $K_c = 37.1$. What is the value for the reverse reaction?

*Solution*

$$K_c(\text{reverse}) = \frac{1}{K_c(\text{forward})} = \frac{1}{37.1} = 0.0269$$

## 4.2 THE MEANING OF $K_C$

Reactions with $K_c$ values significantly greater than 1 (e.g., on the order of $10^4$ or larger) approach completion, and their equilibria lie far to the right or the product side. In general, reactions that go to completion have exceptionally large or undefined $K_c$ values. In contrast, reactions that have $K_c$ values much less than 1 (e.g., on the order of $10^{-4}$ or smaller), as in the case of the $K_a$ value of acetic acid discussed later in this chapter (see Section 4.3), do not undergo significant change. Their equilibria lie far to the left or the reactant side, and thus very little product forms. In the case of acids, the smaller the value of $K_a$, the weaker the acid. For reactions that have equilibrium constant values between 100 and 0.01, it can be said that significant concentrations of both reactants and products are present at equilibrium.

## 4.3 CALCULATIONS FOR $K_C$

A remarkable feature of the equilibrium constant expression is that $K_c$ value for a given reaction at a given temperature is always constant and unique, regardless of which of the three approaches or pathways to equilibrium is followed:

- **Only** reactants present initially.
- **Only** products present initially.
- **Both** reactants **and** products present initially in a random ratio.

The $K_c$ expression format, together with knowledge of the approach or pathway to equilibrium, can prove very valuable in developing a strategy to solving equilibrium problems, as shown in Examples 4.3 and 4.4.

**Examples 4.3:** Expressing and Evaluating $K_c$ and $K_p$

The reaction below is carried out in a 5-L vessel at 600 K. It is one way to convert carbon monoxide into carbon dioxide:

$$CO(g) + H_2O(g) \rightleftharpoons CO_2(g) + H_2(g)$$

At equilibrium, it is found that 0.0200 mol of CO, 0.0215 mol of $H_2O$, 0.0700 mol of $CO_2$, and 2.00 mol of $H_2$ are present. For this reaction, calculate (A) $K_c$ and (B) $K_p$.

*Solution*

A. The reaction is balanced as it stands. All four substances are in the gaseous state and thus have concentrations that can be inserted into the $K_c$ expression.

Since the *molar* amounts are given *at equilibrium*, each is divided by the volume to convert it into molarity and then substituted into the $K_c$ expression. Thus,

$$K_c = \frac{[CO_2][H_2]}{[CO][H_2O]}$$

$$= \frac{(0.0700 \, \text{mol}/5.0 \text{ L})(2.00 \, \text{mol}/ \, 5.0 \text{L})}{(0.0200 \text{ mol}/ \, 5.0 \text{L})(0.0215 \, \text{mol}/ \, 5.0 \text{L})}$$

$$= 326$$

B. To convert to $K_p$, the conversion formula given in Equation 4.4 is used:

$$K_p = K_c (RT)^{\Delta n}$$

Note that in this case, since

$$\Delta n = (1+1) - (1+1) = 0$$

$$K_p = K_c$$

In fact, $K_p = K_c$ whenever the product side and reactant side contain identical numbers of total moles.

**Examples 4.4:** Expressing and Evaluating $K_c$ and $K_p$

Ammonia gas, $NH_3(g)$, is introduced into a previously evacuated reaction vessel in such a way that its initial concentration is 0.500 M. The ammonia decomposes into nitrogen and hydrogen gases, according to the reaction given below, and eventually reaches equilibrium. The equilibrium concentration of nitrogen is found to be 0.116 M. Determine the value of $K_c$.

$$2NH_3(g) \rightleftharpoons N_2(g) + 3H_2(g)$$

## Solution

Knowing the approach or pathway to equilibrium is vital to solving this problem. It is clear in the statement that only reactant is present initially; therefore, all of the products are formed from the decomposition of $NH_3$. The equilibrium concentration of $N_2$ is given as 0.116 M. Because of the 3 to 1 stoichiometric ratio between $H_2$ and $N_2$, 3(0.116 M) = 0.348 M of $H_2$ must have also formed and is present at equilibrium.

Also, because of the stoichiometric ratio between $NH_3$ and $N_2$, 2(0.116 M) = 0.232 M of $NH_3$ must have decomposed. Thus, the concentration of $NH_3$ remaining and present at equilibrium is 0.500 M – 0.232 M = 0.268 M. Therefore, the values for the three species *at equilibrium* are as follows:

$$[NH_3] = 0.268\,M$$

$$[N_2] = 0.116\,M$$

$$[H_2] = 0.348\,M$$

Substitute these values in the $K_c$ expression:

$$K_c = \frac{[N_2][H_2]^3}{[NH_3]^2}$$

$$= \frac{(0.116\,M)(0.348\,M)^3}{(0.269\,M)^2}$$

$$= 0.0681\,M^2$$

## 4.4 PREDICTING EQUILIBRIUM USING A TEST QUOTIENT

It is often useful to know whether a reaction reached equilibrium and, if so, under what conditions. This information can be obtained by measuring the concentrations of all species in the reaction and then determining a test quotient, $Q$. $Q$ is identical in form to $K_c$, except that the values substituted in the expression may or may not be equilibrium concentrations. The criteria for equilibrium are as follows:

- If $Q = K_c$, the reaction is at equilibrium.
- If $Q < K_c$, the reaction is not at equilibrium and needs to proceed to the right.
- If $Q > K_c$, the reaction is not at equilibrium and needs to proceed to the left.

**Example 4.5:** (A & B) Criteria for Reaching Equilibrium

Consider the reaction given in Example 4.2 along with the following concentrations:

$$[SO_2] = 0.054\,M$$

$$[O_2] = 0.020\,M$$

$$[SO_3] = 0.012\,M$$

A. Is this reaction at equilibrium?
B. If not, in which direction will it proceed?

### Solution

A. Write the expression for the test quotient $Q$, then substitute each of the three concentration values given into the expression:

$$Q = \frac{[SO_3]^2}{[SO_2]^2[O_2]} = \frac{(0.012\,M)^2}{(0.054\,M)^2(0.020\,M)}$$

$$= 2.47$$

Since $K_c = 37.1$, $Q$ does not equal $K_c$.
    Thus, the reaction is not at equilibrium.
B. Since $Q < K_c$, the reaction must proceed to the right to reach equilibrium.

## 4.5   STRESSES AND LE CHATELIER'S PRINCIPLE

Once a reaction has reached equilibrium, any external stress that is imposed results in a shift of the equilibrium and a subsequent change in the concentrations of the reactants and products involved. The direction of the shift—right (or product side) or left (or reactant side)—is in accordance with **Le Chatelier's Principle**, which states: If a stress is imposed on a reaction at equilibrium, the reaction will shift either right or left in the way or direction that accommodates or relieves the stress.
    There are five categories of stress:

- The reactant concentration or product concentration is increased or decreased.
- The temperature of the reaction is increased or decreased.
- The volume of the reaction vessel is increased or decreased.
- The pressure of the reaction is increased or decreased (in any one of three different ways).
- A catalyst is added to the reaction.

**Example 4.6:** (A–F) Stresses and LeChatelier's Principle

Consider the following reaction already at equilibrium:

$$CS_2(g) + 3Cl_2(g) \rightleftharpoons S_2Cl_2(g) + CCl_4(g)$$

The enthalpy or heat of reaction is $\Delta H° = -84.3\,kJ$.

Predict what will happen—increase, decrease, or no change—to the equilibrium concentration of $CCl_4$ when each of the following stresses is imposed:

A. $Cl_2$ is added.
B. $S_2Cl_2$ is removed.
C. The temperature is increased.
D. The volume of the reaction vessel is decreased.
E. The pressure is decreased by increasing the volume of the reaction vessel.
F. A catalyst is added.

## Solution

Use Le Chatelier's Principle to answer each part.

A. Adding $Cl_2$ will cause the reaction to shift to the right, increasing $[CCl_4]$.
B. Removing $S_2Cl_2$ will cause the reaction to shift to the right, increasing $[CCl_4]$.
C. To answer this part, the meaning of $\Delta H°$ must be understood. This is the amount of heat energy either liberated or absorbed in the reaction and may be expressed in either kilojoules (kJ) or kilocalories (kcal). If heat energy is liberated, the reaction is exothermic, and the sign of $\Delta H$ is negative by convention. If heat energy is absorbed, the reaction is endothermic, and the sign of $\Delta H$ is positive. Here, the sign is negative, so heat is produced. Hence, since raising the temperature adds heat and heat behaves as a product, increasing the temperature will shift the reaction to the left, decreasing $[CCl_4]$.
D. In the balanced reaction, four molecules of reactant are reacting to form two molecules of product. Increasing the volume of the vessel favors the reactant or left side of the reaction, while decreasing the volume favors the product side. Hence, decreasing the volume will cause the reaction to shift to the right, increasing $[CCl_4]$. In the special case where the two sides of a reaction contain equal numbers of molecules, changing the volume has no effect.
E. If the pressure is decreased by increasing the volume, the effect can be evaluated as a volume stress. As discussed in part D, increasing the volume will cause the reaction to shift to the left and decrease $[CCl_4]$.
   Note, however, that, if the pressure is increased or decreased by the addition of a reactant or product, the effect should be evaluated as a stress discussed in part **A** or **B**.
   Also, note that if the pressure is increased by the addition of an inert or unreactive gas, there is no effect on the equilibrium.
F. A catalyst speeds up the rate of reaction. Since the reaction has already reached equilibrium, adding a catalyst will have no effect.

## 4.6 DEPENDENCE OF $K_C$ ON TEMPERATURE

The exact dependence of $K_c$ (or $K_p$) on temperature is given by the **van't Hoff equation**:

$$\ln\frac{K_c(2)}{K_c(1)} = \frac{\Delta H_{RX}}{R}\left(\frac{1}{T_1} - \frac{1}{T_2}\right)$$

(4.5)

where $K_c(1)$ is the equilibrium constant value at absolute temperature $T_1$ and $K_c(2)$ is the value at $T_2$, $\Delta H_{RX}$ is the enthalpy (or heat) of reaction, and $R$ is the gas constant in units of either joules per mole or calories per mole, consistent with the units of $\Delta H_{RX}$. The similarity of the van't Hoff equation to the Clausius–Clapeyron equation, given as Equation 1.18, is noteworthy.

## 4.7   CALCULATIONS INVOLVING WEAK ACIDS AND BASES

In contrast to strong acids and bases, weak acids and bases dissociate and ionize into $H^+$ ions and $OH^-$ ions, respectively, but to a very limited extent—normally less than 10% and more commonly, less than 5%. The dissociation of weak acids and bases is another class or category of chemical equilibria. To be specific, for every 100 molecules of acid initially present, typically fewer than five molecules dissociate and ionize into hydrogen ions. Dissociate means "separate," and ionize means "form ions." Note that in gas-phase equilibrium reactions, reactants generally undergo only dissociation to form products. In solution-phase equilibrium reactions, as in weak acid or weak base equilibria, reactants undergo not only dissociation but also ionization. Weak acids produce few hydrogen ions, while weak bases produce few hydroxide ions. Their reactions can be characterized by an equilibrium acid ($a$) or base ($b$) constant, i.e., $K_a$ or $K_b$, respectively.

For example, acetic acid (found in vinegar) is a weak acid. Its equilibrium is represented as follows:

$$CH_3COOH(aq) + H_2O(l) \rightleftharpoons H_3O^+(aq) + CH_3COO^-(aq) \qquad (4.6)$$

Here, $H_3O^+$ is the active acid species, and $CH_3COO^-$ is called the **conjugate base** of acetic acid. Recall that the double arrows indicate an equilibrium process, meaning that this reaction does not go to completion and that all four chemical species in the reaction are present at any given time in varying concentrations.

The equilibrium acid constant expression, $K_a$, is as follows:

$$K_a = \frac{\left[H_3O^+\right]\left[CH_3COO^-\right]}{\left[CH_3COOH\right]} \qquad (4.7)$$

The $K_a$ value for acetic acid at room temperature is $1.8 \times 10^{-5}$.

The average percent ionization, depending on initial concentration of the parent acid and temperature, is about 3%–5%. Generally speaking, the smaller the value of the $K_a$, the weaker the acid. Table 4.1(A) gives $K_a$ values for a number of weak acids at 25°C. It is evident then, from comparing $K_a$ values of acetic acid with nitrous acid, that acetic acid is weaker than nitrous acid. Phenol, in turn, is much weaker than either nitrous acid or acetic acid. Polyprotic acids have more than one hydrogen ion to dissociate and ionize and so have more than one dissociation constant, i.e., $K_{a1}$, $K_{a2}$, and $K_{a3}$, which get successively smaller, indicating progressive weakness in acidity. For example, oxalic acid has two $K_a$ values, while citric acid has three.

Note that $K_a$ values can also be expressed as $pK_a$ values, just as hydrogen ion concentrations can be expressed as pH values, simply by taking the negative logarithm of the $K_a$ value. See Equation 1.22 in Chapter 1. Many reference manuals compile weak acid dissociation constants as $pK_a$ values.

## TABLE 4.1
## Equilibrium or Dissociation Constants for (A) Weak Acids $K_a$ (Monoprotic and Polyprotic) and (B) Weak Bases $K_b$

| Monoprotic Acids | | | $K_a$ |
|---|---|---|---|
| $HC_2O_2Cl_3$ | Trichloroacetic acid | $(Cl_3CCO_2H)$ | $2.2 \times 10^{-1}$ |
| $HIO_3$ | Iodic acid | . | $1.69 \times 10^{-1}$ |
| $HC_2HO_2Cl_2$ | Dichloracetic acid | $(Cl_2CHCO_2H)$ | $5.0 \times 10^{-2}$ |
| $HC_2H_2O_2Cl$ | Chloroacetic acid | $(ClH_2CCO_2H)$ | $1.36 \times 10^{-3}$ |
| $HNO_2$ | Nitrous acid | | $7.1 \times 10^{-4}$ |
| $HF$ | Hydrofluoric acid | | $6.8 \times 10^{-4}$ |
| $HOCN$ | Cyanic acid | | $3.5 \times 10^{-4}$ |
| $HCHN_2$ | Formic acid | $(HCO_2H)$ | $1.8 \times 10^{-4}$ |
| $HC_3H_5O_3$ | Lactic acid | $(CH_3CH(OH)CO_2H)$ | $1.38 \times 10^{-4}$ |
| $HC_4H_3N_2O_3$ | Barbituric acid | | $9.8 \times 10^{-5}$ |
| $HC_7H_5O_2$ | Benzoic acid | $(C_6H_5CO_2H)$ | $6.28 \times 10^{-5}$ |
| $HC_4H_7O_2$ | Butanoic acid | $(CH_3CH_2CH_2CO_2H)$ | $1.52 \times 10^{-5}$ |
| $HN_3$ | Hydrazoic acid | | $1.8 \times 10^{-5}$ |
| $HC_2H_3O_2$ | Acetic acid | $(CH_3CO_2H)$ | $1.8 \times 10^{-5}$ |
| $HC_3H_5O_2$ | Propanoic acid | $(CH_3CH_2CO_2H)$ | $1.34 \times 10^{-5}$ |
| $HOCl$ | Hypochlorous acid | | $3.0 \times 10^{-8}$ |
| $HOBr$ | Hypobromous acid | | $2.1 \times 10^{-9}$ |
| $HCN$ | Hydrocyanic acid | | $6.2 \times 10^{-10}$ |
| $HC_6H_5O$ | Phenol | | $1.3 \times 10^{-10}$ |
| $HOI$ | Hypoiodous acid | | $2.3 \times 10^{-11}$ |
| $H_2O_2$ | Hydrogen peroxide | | $1.8 \times 10^{-12}$ |

| Polyprotic Acids | | $K_{a1}$ | $K_{a2}$ | $K_{a3}$ |
|---|---|---|---|---|
| $H_2SO_4$ | Sulfuric acid | Large | $1.0 \times 10^{-2}$ | |
| $H_2CrO_4$ | Chromic acid | 5.0 | $1.5 \times 10^{-6}$ | |
| $H_2C_2O_4$ | Oxalic acid | $5.6 \times 10^{-2}$ | $5.4 \times 10^{-5}$ | |
| $H_3PO_3$ | Phosphorous acid | $3 \times 10^{-2}$ | $1.6 \times 10^{-7}$ | |
| $H_2SO_3$ | Sulfurous acid | $1.2 \times 10^{-2}$ | $6.6 \times 10^{-8}$ | |
| $H_2SeO_3$ | Selenous acid | $4.5 \times 10^{-3}$ | $1.1 \times 10^{-8}$ | |
| $H_2TeO_3$ | Tellurous acid | $3.3 \times 10^{-3}$ | $2.0 \times 10^{-8}$ | |
| $H_2C_3H_2O_4$ | Malonic acid $(HO_2CCH_2CO_2H)$ | $1.4 \times 10^{-3}$ | $2.0 \times 10^{-6}$ | |
| $H_2C_8H_4O_4$ | Phthalic acid | $1.1 \times 10^{-3}$ | $3.9 \times 10^{-6}$ | |
| $H_2C_4H_4O_6$ | Tartaric acid | $9.2 \times 10^{-4}$ | $4.3 \times 10^{-5}$ | |
| $H_2C_6H_6O_6$ | Ascorbic acid | $7.9 \times 10^{-5}$ | $1.6 \times 10^{-12}$ | |
| $H_2CO_3$ | Carbonic acid | $4.5 \times 10^{-7}$ | $4.7 \times 10^{-11}$ | |
| $H_3PO_4$ | Phosphoric acid | $7.1 \times 10^{-3}$ | $6.3 \times 10^{-8}$ | $4.5 \times 10^{-13}$ |
| $H_3AsO_4$ | Arsenic acid | $5.6 \times 10^{-3}$ | $1.7 \times 10^{-7}$ | $4.0 \times 10^{-12}$ |
| $H_3C_6H_5O_7$ | Citric acid | $7.1 \times 10^{-4}$ | $1.7 \times 10^{-5}$ | $4.0 \times 10^{-7}$ |

*(Continued)*

**TABLE 4.1 (*Continued*)**

**Equilibrium or Dissociation Constants for (A) Weak Acids $K_a$ (Monoprotic and Polyprotic) and (B) Weak Bases $K_b$**

|  | Weak Bases | $K_b$ |
|---|---|---|
| $(CH_3)_2NH$ | Dimethylamine | $9.6 \times 10^{-4}$ |
| $CH_3NH_2$ | Methylamine | $4.4 \times 10^{-4}$ |
| $CH_3CH_2NH_2$ | Ethylamine | $4.3 \times 10^{-4}$ |
| $(CH_3)_3N$ | Trimethylamine | $7.4 \times 10^{-5}$ |
| $NH_3$ | Ammonia | $1.8 \times 10^{-5}$ |
| $N_2H_4$ | Hydrazine | $8.5 \times 10^{-7}$ |
| $NH_2OH$ | Hydroxylamine | $6.6 \times 10^{-9}$ |
| $C_5H_5N$ | Pyridine | $1.5 \times 10^{-9}$ |
| $C_6H_5NH_2$ | Aniline | $4.1 \times 10^{-10}$ |
| $PH_3$ | Phosphine | $10^{-28}$ |

There are two ways to compute [H⁺] or [H₃O⁺] for a solution of a weak acid. If the percent ionization is given, multiply this by the initial concentration of the acid to obtain [H⁺]. See Example 4.7. If the $K_a$ value is given along with the *initial* concentration of the acid, Equation 4.7 can be rearranged and solved for [H₃O⁺]. See Example 4.8.

**Example 4.7:** Determining the H⁺ concentration and pH of Two Weak Acids

A solution of 0.14 M nitrous acid, $HNO_2$, is 5.7% ionized. Calculate [H⁺].

*Solution*

Nitrous acid is obviously a weak acid since its percent ionization is given as less than 10% and certainly much less than 100%. Thus,

$$\left[H^+\right] = 5.7\% \times 0.14\ M = 7.98 \times 10^{-3}\ M$$

**Example 4.8:** Determining the H⁺ concentration and pH of Two Weak Acids

$K_a$ for formic acid, HCOOH, contained in ant venom, is $1.80 \times 10^{-4}$. Find the pH of a solution that initially contains 0.200 mol of HCOOH in 500 mL of solution.

*Solution*

Since formic acid has a $K_a$ value significantly less than 1.0, it must be a weak acid. Its equilibrium equation and $K_a$ expression are as follows:

$$HCOOH(aq) + H_2O(l) \rightleftharpoons H_3O^+(aq) + HCOO^-(aq)$$

$$K_a = \frac{\left[H_3O^+\right]\left[HCOO^-\right]}{HCOOH} = 1.8 \times 10^{-4}$$

The strategy used in solving this type of problem is one that is commonly used in solving other equilibrium problems. The above expression for $K_a$ equated to its value is a useful equation; it can be solved for $[H_3O^+]$, which can then be converted into pH.

The *initial* concentration of HCOOH is given:

$$[HCOOH] = \frac{0.20\,mol}{0.500\,L} = 0.400\,M$$

However, all concentrations inserted into the $K_a$ expression must be equilibrium or final values. Therefore, let $x =$ the number of moles per liter of HCOOH that dissociate and ionize into products. Since 1 mol of HCOOH produces 1 mol of $H_3O^+$ *and* 1 mol of HCOO$^-$, and all products come from only one reactant, *at equilibrium*, then:

$$\left[H_3O^+\right] = x,$$

$$\left[HCOO^-\right] = x$$

$$[HCOOH] = 0.40 \;-\; x$$

Substitute these values in the $K_a$ expression, Equation 4.7:

$$\frac{(x)(x)}{0.40 - x} = 1.8 \times 10^{-4}$$

This is a quadratic equation in $x$. A simplifying assumption may be made. Since $x \ll 0.40$, it may be neglected in the denominator, reducing this to an easier-to-solve abridged quadratic equation. Then,

$$x^2 = \left(1.8 \times 10^{-4}\right)(0.40)$$

$$x = 8.49 \times 10^{-3}\,M$$

Thus,

$$\left[H_3O^+\right] = 8.49 \times 10^{-3}\,M$$

Finally:

$$pH = -\log\left[H_3\,O^+\right] = -\log(8.49 \times M) = 2.07$$

*Note:* A good rule of thumb to follow to determine whether the simplifying assumption is justified is as follows: Let $M_a =$ initial molarity of acid; then,

If $M_a/K_a > 100$, then $M_a - x \cong M_a$ is a good approximation.

The error incurred is usually about 1%, which is generally acceptable. The $M_a/K_a$ ratio in the example above produces a value of about 2,200. If the complete or unabridged quadratic equation is solved for this example, the calculated value for the equilibrium concentration of $H_3O^+$ is $8.57 \times 10^{-3}$ M. Thus, the error incurred is about 0.93% or slightly less than 1%.

## 4.8  HYDROLYSIS

Some compounds that appear neutral actually have acidic or alkaline properties. For example, when sodium acetate, $NaCH_3COO$, is dissolved in water, it reacts with water to form acetic acid (a weak acid) and a hydroxide ion. This process is called **hydrolysis**, and the resulting solution is slightly alkaline. Remember that the acetate ion is the conjugate base of acetic acid. Other basic salts in this category include sodium cyanide, potassium nitrite, and lithium carbonate.

Similarly, ammonium chloride, $NH_4Cl$, produces a slightly acidic solution when dissolved in water, again because of hydrolysis. Ammonium ion is the **conjugate acid** of ammonium hydroxide, and ammonia, $NH_3$, is the **conjugate base** of $NH_4^+$. Another salt in this category is aluminum chloride, $AlCl_3$. Gases such as $CO_2$ and $SO_2$ also undergo hydrolysis and produce $H^+$ ions and $HCO_3^-$ and $HSO_3^-$ ions, respectively.

**Example 4.9:** Hydrolysis of a Weak Salt

Calculate the pH of a .20 M solution of ammonium chloride, $NH_4Cl$.

### Solution

$NH_4Cl$ is the salt of a weak base, $NH_3$, and hence hydrolyzes when dissolved in water to form a slightly acidic solution.

$$NH_4^+(aq) + H_2O(l) \rightleftharpoons H_3O^+(aq) + NH_3(aq)$$

The hydrolysis constant expression, $K_h$, can be expressed as follows:

$$K_h = \frac{[H_3O^+][NH_3]}{[NH_4^+]}$$

The value for $K_h$ is not given but can be found by applying the relationship

$$K_h = \frac{K_w}{K_b} = \frac{1.0 \times 10^{-14}}{1.8 \times 10^{-5}} = 5.56 \times 10^{-10}$$

where $K_b$ is the value of the equilibrium base constant for $NH_3$ in $H_2O$ and is found in Table 4.1B. It is analogous to $K_a$:

$$NH_3 + H_2O \rightleftharpoons NH_4^+ + OH^-$$

The same strategy is then used as in Example 4.8. At *equilibrium*, then,

$$[H_3O^+] = x$$

$$[NH_3] = x$$

$$[NH_4^+] = 0.20 - x$$

Therefore,

$$5.56 \times 10^{-10} = \frac{(x)(x)}{0.20 - x}$$

Use the same simplifying assumption as in Example 4.8, assume $x \ll 0.20$ M, and solve for $x$:

$$x^2 = (0.20\,\text{M})(5.56 \times 10^{-10})$$

$$x = 1.1 \times 10^{-5}$$

Thus,

$$[H_3O^+] = 1.1 \times 10^{-5}\,\text{M}$$

And,

$$pH = -\log[H_3O^+] = -\log(1.1 \times 10^{-5}\,\text{M}) = 4.96$$

The pH is less than 7.0, reflecting an acidic solution, as expected.

   Note: For the hydrolysis of an ion producing a *basic or alkaline* solution, $K_h = K_w/K_a$.

## 4.9   BUFFER SOLUTIONS AND THE HENDERSON–HASSELBALCH EQUATION

Buffer solutions are solutions that protect against large shifts in pH in the event of a shock due to the sudden addition of a strong acid or base. The pH of a buffer solution does, however, change slightly. Such a solution can be made with a combination of either of the following:

- A weak acid and its salt (conjugate base)
- A weak base and its salt (conjugate acid)

Thus, acetic acid plus sodium acetate constitutes a buffer system, as does ammonium hydroxide plus ammonium chloride.

   To determine the pH of a buffer solution, the **Henderson–Hasselbalch equation** may be used. In the case of a weak acid (e.g., $CH_3COOH$) and its conjugate base salt (e.g., $NaCH_3COO$):

$$pH = pK_a + \log\frac{\left[\text{conjugate base}\right]}{\left[\text{weak acid}\right]} \tag{4.8}$$

The buffering capacity of a system refers to the amount of acid or base it can absorb before its pH changes. Buffering capacity is generally at a maximum when [weak acid] = [conjugate base], so that pH = $pK_a$.

## 4.10 AMPHOTERISM

**Amphoterism** refers to the property of certain compounds to act as both acids and bases. For example, the compound $Al(OH)_3$, which is categorically a base, may react with acid to form $[Al (H_2O)_6]^{3+}$ *and* with excess base to form $[Al(OH)_4]^{-1}$. Both of these species are termed **complex ions** and are water-soluble. They are governed by complex-ion equilibrium constants, as explained in Section 4.16 of this chapter. Aluminum oxide, $Al_2O_3$, is also amphoteric. The hydroxides, as well as oxides of zinc(II) and chromium(III), are similarly amphoteric.

## 4.11 SOLUBILITY PRODUCT CONSTANTS

Like the acids and bases discussed in the previous section, another group of substances comprised generally of ionic compounds that have very limited solubility in water deserves separate treatment. Examples include magnesium hydroxide (commonly known as milk of magnesia, an antacid), barium sulfate (used in enemas to diagnose colonic tumors), and silver chloride. These substances are generally referred to as "insoluble" or "slightly soluble" and are governed by a solubility equilibrium expression and a solubility product constant known as $K_{sp}$, identical in concept to the principle of equilibrium discussed in Section 4.3.

The solubility of any salt depends on temperature. With few exceptions, the solubility of most salts increases with increasing temperature, usually in a nonlinear relationship. See Table 4.2 for a select list of salts that are considered "freely soluble" and their solubilities as a function of temperature. A more complete list is available in the *Handbook of Chemistry and Physics*, CRC Press.

When a salt has dissolved to the maximum extent in a given volume of water at a given temperature, the solution is said to be **saturated**. Until that point is reached, the solution is **unsaturated**. Of course, this depends on temperature. If a saturated solution is heated above room temperature, more salt can be dissolved and be held in solution. If this solution is later slowly cooled (usually called *reversibly cooled*), it is possible to cool it to back down to room temperature (or below), such that all of the dissolved salt can exist in the solution phase without precipitating out. Once this lower temperature is reached, the solution is said to be **supersaturated**. A sudden shock,

**TABLE 4.2**
**Solubility of Select Salts As a Function of Temperature (in grams/100 mL)**

| Salt | Formula | 0°C | 20°C | 50°C | 70°C | 100°C |
|------|---------|-----|------|------|------|-------|
| Aluminum nitrate | $Al(NO_3)_3$ | 60 | 73.9 | 96 | 120 | 160 |
| Aluminum sulfate | $Al_2(SO_4)_3$ | 31.2 | 36.4 | 52.2 | 66.2 | 89.0 |
| Copper (II) nitrate | $Cu(NO_3)_2$ | 83.5 | 156 | 182 | 208 | 247 |
| Magnesium nitrate | $Mg(NO_3)_2$ | 62.1 | 73.6 | 80.1 | — | — |
| Sodium chloride | $NaCl$ | 35.6 | 36.1 | 36.7 | 37.9 | 39.0 |

e.g., mechanical disturbance, however, can cause instantaneous precipitation and the corresponding release of energy in the form of heat. The solubility of a salt is commonly measured in units of moles per liter, grams per liter, or milligrams per liter (often referred to as *parts per million* or simply *ppm*).

## 4.12 DEFINITION OF SOLUBILITY PRODUCT CONSTANT $K_{SP}$

Consider, for example, the slightly soluble salt $CaF_2(s)$ in equilibrium with water. Its limited dissociation and ionization in water can be expressed as follows:

$$CaF_2(s) \rightleftharpoons Ca^{2+}(aq) + 2F^-(aq) \qquad (4.9)$$

And its **solubility product constant**, $K_{sp}$, as

$$K_{sp} = \left[ Ca^{2+} \right]\left[ F^- \right]^2 \qquad (4.10)$$

Note that $[CaF_2(s)]$ does not appear in the $K_{sp}$ expression since it represents the undissolved portion, and it is pure solid. It has no concentration.

Also note that $[Ca^{2+}]$ represents the concentration of $Ca^{2+}$ ion dissolved in water, while $[F^-]$ represents the $F^-$ ion concentration dissolved in water.

Note also that when a given amount of $CaF_2(s)$ is dissolved in a known amount of water:

$$\left[ Ca^{2+}(aq) \right] = 2\left[ F^-(aq) \right] = \left[ CaF_2(aq) \right]$$

where $CaF_2(aq)$ is the portion that is dissolved in water. In other words, whatever the proportion of $CaF_2(s)$ that dissolves in water, the same molar concentration is present as $Ca^{2+}$ (aq) ion, but twice that molar concentration is present as $F^-$ (aq) ion.

The smaller the $K_{sp}$ value, the lower the solubility of the salt. It follows that the $K_{sp}$ value of a salt can be calculated by determining the solubility of each ion and then using Equation 4.10.

## 4.13 CALCULATING THE MOLAR SOLUBILITY FROM $K_{SP}$

The molar solubility of any slightly soluble salt, along with the concentration of any of its ions, can be calculated from its $K_{sp}$ value. The formula of the salt must be known, however, so that its dissociation and ionization can be written.

It is important to note that simply comparing the $K_{sp}$ values of two salts to determine which has a higher or lower molar solubility can often be misleading. This comparison of numerical values is valid only when the two salts have the same stoichiometry, that is, the subscripts in the formulas are identical.

**Example 4.10:** (A & B) Computing the Molar Solubility of an Insoluble Salt

Calculate:

A. The molar solubility of calcium fluoride, $CaF_2$, and
B. The concentration of the fluoride ion, in solution at 25°C. The $K_{sp}$ of $CaF_2$ is $4.0 \times 10^{-11}$ at 25°C. Here, the unit of $K_{sp}$ is $(moles/L)^3$ or $M^3$. This unit may change depending on the identity, i.e., the subscripts of the ions, of the salt.

*Solution*

A. The equilibrium for the salt is as follows:

$$CaF_2(s) \rightleftharpoons Ca^{2+}(aq) + 2F^-(aq)$$

The $K_{sp}$ expression is

$$K_{sp} = \left[Ca^{2+}\right]\left[F^-\right]^2 = 4.0 \times 10^{-11} \, M^3$$

Let $S$ = the solubility (saturation concentration) of the $Ca^{2+}$ ion in moles per liter.
    Then, $2S$ = the solubility (saturation concentration) of the $F^-$ ion.
    Substitute $S$ into the $K_{sp}$ expression:

$$(S)(2S)^2 = 4.0 \times 10^{-11}$$

$$4S^3 = 4.0 \times 10^{-11}$$

$$S = 2.15 \times 10^{-4} \, M$$

Thus, the molar solubility of a $CaF_2$ is $2.15 \times 10^{-4}$ M. This is also $[Ca^{2+}]$.
B. $[F^-] = 2 \times [Ca^{2+}] = 2 \times 2.15 \times 10^{-4}$ M $= 4.30 \times 10^{-4}$ M.

**Example 4.11:** Calculating the $K_{sp}$ of an Insoluble Salt

The molar solubility of tin iodide, $SnI_2$, is $1.28 \times 10^{-2}$ mol/L. What is $K_{sp}$ for this compound?

*Solution*
The solubility equilibrium for $SnI_2$ is

$$SnI_2(s) \rightleftharpoons Sn^{2+}(aq) + 2I^-(aq)$$

The $K_{sp}$ expression is

$$K_{sp} = \left[Sn^{2+}\right]\left[I^-\right]^2$$

Note that 1.0 mol of $SnI_2$ produces 1.0 mol of $Sn^{2+}$, but 2.0 mol of $I^-$.

$$\left[Sn^{2+}\right] = 1.28 \times 10^{-2}\,M$$

$$\left[I^-\right] = (2) \times 1.28 \times 10^{-2}\,M = 2.56 \times 10^{-2}\,M$$

Substituting these values into the $K_{sp}$ expression yields:

$$K_{sp} = \left(1.28 \times 10^{-2}\,M\right)\left(2.56 \times 10^{-2}\,M\right)^2 = 8.4 \times 10^{-6}\,M^3$$

## 4.14  COMMON AND UNCOMMON ION AND pH EFFECTS

Occasionally, the solvent in which a salt is to be dissolved is not pure water but rather contains an ion in common with the salt to be dissolved. The common ion may be a cation or an anion, such as $Ca^{2+}$ or $F^-$. The common ion may also be the $H^+$ ion or $OH^-$ ion, which manifests itself in a pH value less than or greater than 7, respectively. This is the **common ion effect**, and the net result is to reduce significantly the solubility of the salt in question.

The presence of a background salt with no ions in common with the salt to be dissolved is known as the **uncommon ion** or **salt effect**. In contrast, this effect only slightly increases the solubility of the salt in question.

In general, most salts are more soluble in hot water than in cold; that is, the solubilities of most salts increase with increasing temperature. In the temperature range of 0–100°C, many common salts demonstrate a nearly linear relationship between solubility and temperature, with varying, positive slopes. Some notable exceptions include $KNO_3$, which shows an exponential relationship, and $Li_2SO_4$, which shows a linear but decreasing solubility with temperature.

**Example 4.12:** Calculating the Molar Solubility With the Common Ion effect

Magnesium hydroxide, $Mg(OH)_2$, is a common antacid and has a $K_{sp} = 1.8 \times 10^{-11}\,M^3$. Its limited solubility is what makes this otherwise strong base suitable as an antacid for human consumption. Find the molar solubility of $Mg(OH)_2$ in a laboratory solution whose pH is 13.12.

### Solution

To find the molar solubility of $Mg(OH)_2$ in distilled water, the method outlined in Example 4.10 would be followed. In this case, however, the solution is basic with an excess of $OH^-$ ions. Since these $OH^-$ ions are in common with the $OH^-$ ions contained in $Mg(OH)_2$, the common ion effect must be taken into account because it will reduce the solubility of $Mg(OH)_2$.

First determine $\left[OH^-\right]$ as follows:

$$pOH = 14.00 - pH = 14.00 - 13.12 = 0.88$$

$$\left[OH^-\right] = 10^{-pOH} = 10^{-0.88} = 1.32 \times 10^{-1} \, M$$

The $K_{sp}$ expression for $Mg(OH)_2$ is:

$$K_{sp} = \left[Mg^{2+}\right]\left[OH^-\right]^2$$

Use the strategy outlined in Example 4.10.

Let $\left[Mg^{2+}\right] = S$.

And $\left[OH^-\right] = 2S + 1.32 \times 10^{-1}$ since the *total* hydroxide ion concentration is the sum of the contribution of hydroxide ion coming from the already alkaline solution, i.e., $1.32 \times 10^{-1}M$, *plus* the contribution of hydroxide ion coming from the $Mg(OH)_2$ source, represented as $2S$, since the hydroxide ion concentration must be twice the magnesium ion concentration, as indicated in the chemical formula of $Mg(OH)_2$.
     Now substitute these values, along with $K_{sp}$ value, in the $K_{sp}$ expression:

$$1.8 \times 10^{-11} = (S)\left(2S + 1.32 \times 10^{-1}\right)^2$$

This is a cubic equation in $S$. To simplify the calculation, assume that $2S$ term is negligible compared with $1.32 \times 10^{-1}$ as an additive term. Then:

$$1.8 \times 10^{-11} = (S)\left(1.32 \times 10^{-1}\right)^2$$

$$S = 1.0 \times 10^{-9} \, M$$

This result represents the solubility of $Mg(OH)_2$.

## 4.15   PREDICTING PRECIPITATION

When two solutions are mixed, it is possible that an insoluble precipitate will form and settle to the bottom. Whether this happens depends on the identities of the ions present in the solutions and their respective concentrations. If the ions for a potentially insoluble precipitate are present, all that needs to be done to predict precipitation is to calculate a hypothetical solubility product constant, $Q$, in the same manner as $K_{sp}$ is calculated. The $Q$ value is then compared with the $K_{sp}$ value, subject to the following criteria or conditions:

- If $Q \leq K_{sp}$, no precipitate forms
- If $Q > K_{sp}$, a precipitate forms.

In general, conditions that favor **completeness of precipitation** are as follows:

- A very small value of $K_{sp}$
- High initial ion concentrations
- A concentration of common ion that considerably exceeds the concentration of the target ion to be precipitated.

### Example 4.13: Predicting Precipitation

Will a precipitate form if 50.0 mL of $1.2 \times 10^{-3}$ M $Pb(NO_3)_2$ is added to 50.0 mL of $2.0 \times 10^{-4}$ M $Na_2SO_4$?

*Solution*
The addition of the two solutions will result in a double-replacement reaction. Two products $NaNO_3$ and $PbSO_4$ will form and are thus candidates as insoluble salts. A check of Table 4.3 reveals that only $PbSO_4$ has a $K_{sp}$ value ($1.6 \times 10^{-8}$) and is therefore a possible precipitate. To confirm whether a precipitate actually forms, evaluate $Q$ as follows:

$$Q = \left[ Pb^{2+}(aq) \right]\left[ SO_4^{2-}(aq) \right]$$

$$\left[ Pb^{2+}(aq) \right] = \frac{(0.050\,L)(1.2 \times 10^{-3}\,M)}{(0.100\,L)}$$

$$= 6.0 \times 10^{-4}\,M$$

$$\left[ SO_4^{2-}(aq) \right] = \frac{(0.050\,L)(2.0 \times 10^{-4}\,M)}{0.100\,L}$$

$$= 1.0 \times 10^{-4}\,M$$

Substitute these values for concentrations into the expression for $Q$:

$$Q = (6.0 \times 10^{-4}\,M)(1.0 \times 10^{-4}\,M)$$

$$= 6.0 \times 10^{-8}\,M$$

But $K_{sp} = 1.6 \times 10^{-8}$.

Since $Q > K_{sp}$, then a precipitate forms!

## TABLE 4.3
### The Solubility Product Constants, $K_{sp}$, of Common Insoluble Salts at 25°C

| Salt | Solubility Equilibrium | $K_{sp}$ |
|---|---|---|
| | **Fluorides** | |
| $MgF_2$ | $MgF_2(s) \rightleftharpoons Mg^{2+}(aq) + 2F^-(aq)$ | $6.6 \times 10^{-9}$ |
| $CaF_2$ | $CaF_2(s) \rightleftharpoons Ca^{2+}(aq) + 2F^-(aq)$ | $3.9 \times 10^{-11}$ |
| $SrF_2$ | $SrF_2(s) \rightleftharpoons Sr^{2+}(aq) + 2F^-(aq)$ | $2.9 \times 10^{-9}$ |
| $BaF_2$ | $BaF_2(s) \rightleftharpoons Ba^{2+}(aq) + 2F^-(aq)$ | $1.7 \times 10^{-6}$ |
| $LiF$ | $LiF(s) \rightleftharpoons Li^+(aq) + F^-(aq)$ | $1.7 \times 10^{-3}$ |
| $PbF_2$ | $PbF_2(s) \rightleftharpoons Pb^{2+}(aq) + 2F^-(aq)$ | $3.6 \times 10^{-8}$ |
| | **Chlorides** | |
| $CuCl$ | $CuCl(s) \rightleftharpoons Cu^+(aq) + Cl^-(aq)$ | $1.9 \times 10^{-7}$ |
| $AgCl$ | $AgCl(s) \rightleftharpoons Ag^+(aq) + Cl^-(aq)$ | $1.8 \times 10^{-10}$ |
| $Hg_2Cl_2$ | $Hg_2Cl_2(s) \rightleftharpoons Hg_2^{2+}(aq) + 2Cl^-(aq)$ | $1.2 \times 10^{-18}$ |
| $TlCl$ | $TlCl(s) \rightleftharpoons Tl^+(aq) + Cl^-(aq)$ | $1.8 \times 10^{-4}$ |
| $PbCl_2$ | $PbCl_2(s) \rightleftharpoons Pb^{2+}(aq) + 2Cl^-(aq)$ | $1.7 \times 10^{-5}$ |
| $AuCl_3$ | $AuCl_3(s) \rightleftharpoons Au^{3+}(aq) + 3Cl^-(aq)$ | $3.2 \times 10^{-25}$ |
| | **Bromides** | |
| $CuBr$ | $CuBr(s) \rightleftharpoons Cu^+(aq) + Br^-(aq)$ | $5 \times 10^{-9}$ |
| $AgBr$ | $AgBr(s) \rightleftharpoons Ag^+(aq) + Br^-(aq)$ | $5.0 \times 10^{-13}$ |
| $Hg_2Br_2$ | $Hg_2^{2+}(s) \rightleftharpoons Hg_2^{2+}(aq) + 2Br^-(aq)$ | $5.6 \times 10^{-13}$ |
| $HgBr_2$ | $HgBr_2(s) \rightleftharpoons Hg^{2+}(aq) + 2Br^-(aq)$ | $1.3 \times 10^{-19}$ |
| $PbBr_2$ | $PbBr_2(s) \rightleftharpoons Pb^{2+}(aq) + 2Br^-(aq)$ | $2.1 \times 10^{-6}$ |
| | **Iodides** | |
| $CuI$ | $CuI(s) \rightleftharpoons Cu^+(aq) + I^-(aq)$ | $1 \times 10^{-12}$ |
| $AgI$ | $AgI(s) \rightleftharpoons Ag^+(aq) + I^-(aq)$ | $8.3 \times 10^{-17}$ |
| $Hg_2I_2$ | $Hg_2I_2(s) \rightleftharpoons Hg_2^{2+}(aq) + 2I^-(aq)$ | $4.7 \times 10^{-29}$ |
| $HgI_2$ | $HgI_2(s) \rightleftharpoons Hg^{2+}(aq) + 2I^-(aq)$ | $1.1 \times 10^{-28}$ |
| $PbI_2$ | $PbI_2(s) \rightleftharpoons Pb^{2+}(aq) + 2I^-(aq)$ | $7.9 \times 10^{-9}$ |
| | **Hydroxides** | |
| $Mg(OH)_2$ | $Mg(OH)_2(s) \rightleftharpoons Mg^{2+}(aq) + 2OH^-(aq)$ | $7.1 \times 10^{-12}$ |
| $Ca(OH)_2$ | $Ca(OH)_2(s) \rightleftharpoons Ca^{2+}(aq) + 2OH^-(aq)$ | $6.5 \times 10^{-6}$ |
| $Mn(OH)_2$ | $Mn(OH)_2(s) \rightleftharpoons Mn^{2+}(aq) + 2OH^-(aq)$ | $1.6 \times 10^{-13}$ |
| $Fe(OH)_2$ | $Fe(OH)_2(s) \rightleftharpoons Fe^{2+}(aq) + 2OH^-(aq)$ | $7.9 \times 10^{-16}$ |
| $Fe(OH)_3$ | $Fe(OH)_3(s) \rightleftharpoons Fe^{3+}(aq) + 3OH^-(aq)$ | $1.6 \times 10^{-39}$ |
| $Co(OH)_2$ | $Co(OH)_2(s) \rightleftharpoons Co^{2+}(aq) + 2OH^-(aq)$ | $1 \times 10^{-15}$ |
| $Co(OH)_3$ | $Co(OH)_3(s) \rightleftharpoons Co^{3+}(aq) + 3OH^-(aq)$ | $3 \times 10^{-45}$ |
| $Ni(OH)_2$ | $Ni(OH)_2(s) \rightleftharpoons Ni^{2+}(aq) + 2OH^-(aq)$ | $6 \times 10^{-16}$ |
| $Cu(OH)_2$ | $Cu(OH)_2(s) \rightleftharpoons Cu^{2+}(aq) + 2OH^-(aq)$ | $4.8 \times 10^{-20}$ |
| $V(OH)_3$ | $V(OH)_3(s) \rightleftharpoons V^{3+}(aq) + 3OH^-(aq)$ | $4 \times 10^{-35}$ |
| $Cr(OH)_3$ | $Cr(OH)_3(s) \rightleftharpoons Cr^{3+}(aq) + 3OH^-(aq)$ | $2 \times 10^{-30}$ |
| $Ag_2O$ | $Ag_2O(s) + H_2O \rightleftharpoons 2Ag^+(aq) + 2OH^-(aq)$ | $1.9 \times 10^{-8}$ |

*(Continued)*

**TABLE 4.3 (*Continued*)**
**The Solubility Product Constants, $K_{sp}$, of Common Insoluble Salts at 25°C**

| Salt | Solubility Equilibrium | $K_{sp}$ |
|---|---|---|
| $Zn(OH)_2$ | $Zn(OH)_2(s) \rightleftharpoons Zn^{2+}(aq) + 2OH^-(aq)$ | $3.0 \times 10^{-16}$ |
| $Cd(OH)_2$ | $Cd(OH)_2(s) \rightleftharpoons Cd^{2+}(aq) + 2OH^-(aq)$ | $5.0 \times 10^{-15}$ |
| $Al(OH)_3$ (alpha form) | $Al(OH)_3(s) \rightleftharpoons Al^{3+}(aq) + 3OH^-(aq)$ | $3 \times 10^{-34}$ |
| | **Cyanides** | |
| $AgCN$ | $AgCN(s) \rightleftharpoons Ag^+(aq) + CN^-(aq)$ | $2.2 \times 10^{-16}$ |
| $Zn(CN)_2$ | $Zn(CN)_2(s) \rightleftharpoons Zn^{2+}(aq) + 2CN^-(aq)$ | $3 \times 10^{-16}$ |
| | **Sulfites** | |
| $CaSO_3$ | $CaSO_3(s) \rightleftharpoons Ca^{2+}(aq) + SO_3^{2-}(aq)$ | $3 \times 10^{-7}$ |
| $Ag_2SO_3$ | $Ag_2SO_3(s) \rightleftharpoons 2Ag^+(aq) + SO_3^{2-}(aq)$ | $1.5 \times 10^{-14}$ |
| $BaSO_3$ | $BaSO_3(s) \rightleftharpoons Ba^{2+}(aq) + SO_3^{2-}(aq)$ | $8 \times 10^{-7}$ |
| $CaSO_4$ | $CaSO_4(s) \rightleftharpoons Ca^{2+}(aq) + SO_4^{2-}(aq)$ | $2.4 \times 10^{-5}$ |
| $SrSO_4$ | $SrSO_4(s) \rightleftharpoons Sr^{2+}(aq) + SO_4^{2-}(aq)$ | $3.2 \times 10^{-7}$ |
| $BaSO_4$ | $BaSO_4(s) \rightleftharpoons Ba^{2+}(aq) + SO_4^{2-}(aq)$ | $1.1 \times 10^{-10}$ |
| $RaSO_4$ | $RaSO_4(s) \rightleftharpoons Ra^{2+}(aq) + SO_4^{2-}(aq)$ | $4.3 \times 10^{-11}$ |
| $Ag_2SO_4$ | $Ag_2SO_4(s) \rightleftharpoons 2Ag^+(aq) + SO_4^{2-}(aq)$ | $1.5 \times 10^{-5}$ |
| $Hg_2SO_4$ | $Hg_2SO_4(s) \rightleftharpoons Hg_2^{2+}(aq) + SO_4^{2-}(aq)$ | $7.4 \times 10^{-7}$ |
| $PbSO_4$ | $PbSO_4(s) \rightleftharpoons Pb^{2+}(aq) + SO_4^{2-}(aq)$ | $1.6 \times 10^{-8}$ |
| | **Chromates** | |
| $BaCrO_4$ | $BaCrO_4(s) \rightleftharpoons Ba^{2+}(aq) + CrO_4^{2-}(aq)$ | $2.1 \times 10^{-10}$ |
| $CuCrO_4$ | $CuCrO_4(s) \rightleftharpoons Ba^{2+}(aq) + CrO_4^{2-}(aq)$ | $3.6 \times 10^{-6}$ |
| $Ag_2CrO_4$ | $Ag_2CrO_4(s) \rightleftharpoons 2Ag^+(aq) + CrO_4^{2-} (aq)$ | $1.2 \times 10^{-12}$ |
| $Hg_2CrO_4$ | $Hg_2CrO_4(s) \rightleftharpoons Hg_2^{2+}(aq) + CrO_4^{2-}(aq)$ | $2.0 \times 10^{-9}$ |
| $CaCrO_4$ | $CaCrO_4(s) \rightleftharpoons Ca^{2+}(aq) + CrO_4^{2-}(aq)$ | $7.1 \times 10^{-4}$ |
| $PbCrO_4$ | $PbCrO_4(s) \rightleftharpoons Pb^{2+}(aq) + CrO_4^{2-}(aq)$ | $1.8 \times 10^{-14}$ |
| | **Carbonates** | |
| $MgCO_3$ | $MgCO_3(s) \rightleftharpoons Mg^{2+}(aq) + CO_3^{2-}(aq)$ | $3.5 \times 10^{-8}$ |
| $CaCO_3$ | $CaCO_3(s) \rightleftharpoons Ca^{2+}(aq) + CO_3^{2-}(aq)$ | $4.5 \times 10^{-9}$ |
| $SrCO_3$ | $SrCO_3(s) \rightleftharpoons Sr^{2+}(aq) + CO_3^{2-}(aq)$ | $9.3 \times 10^{-10}$ |
| $BaCO_3$ | $BaCO_3(s) \rightleftharpoons Ba^{2+}(aq) + CO_3^{2-}(aq)$ | $5.0 \times 10^{-9}$ |
| $MnCO_3$ | $MnCO_3(s) \rightleftharpoons Mn^{2+}(aq) + CO_3^{2-}(aq)$ | $5.0 \times 10^{-10}$ |
| $FeCO_3$ | $FeCO_3(s) \rightleftharpoons Fe^{2+}(aq) + CO_3^{2-}(aq)$ | $2.1 \times 10^{-11}$ |
| $CoCO_3$ | $CoCO_3(s) \rightleftharpoons Co^{2+}(aq) + CO_3^{2-}(aq)$ | $1.0 \times 10^{-10}$ |
| $NiCO_3$ | $NiCO_3(s) \rightleftharpoons Ni^{2+}(aq) + CO_3^{2-}(aq)$ | $1.3 \times 10^{-7}$ |
| $CuCO_3$ | $CuCO_3(s) \rightleftharpoons Cu^{2+}(aq) + CO_3^{2-}(aq)$ | $2.3 \times 10^{-10}$ |
| $Ag_2CO_3$ | $Ag_2CO_3(s) \rightleftharpoons 2Ag^+(aq) + CO_3^{2-}(aq)$ | $8.1 \times 10^{-12}$ |
| $Hg_2CO_3$ | $Hg_2CO_3(s) \rightleftharpoons 2Hg^+(aq) + CO_3^{2-}(aq)$ | $8.9 \times 10^{-17}$ |
| $ZnCO_3$ | $ZnCO_3(s) \rightleftharpoons Zn^{2+}(aq) + CO_3^{2-}(aq)$ | $1.0 \times 10^{-10}$ |
| $CdCO_3$ | $CdCO_3(s) \rightleftharpoons Cd^{2+}(aq) + CO_3^{2-}(aq)$ | $1.8 \times 10^{-14}$ |
| $PbCO_3$ | $PbCO_3(s) \rightleftharpoons Pb^{2+}(aq) + CO_3^{2-}(aq)$ | $7.4 \times 10^{-14}$ |

*(Continued)*

**TABLE 4.3 (*Continued*)**
**The Solubility Product Constants, $K_{sp}$, of Common Insoluble Salts at 25°C**

| Salt | Solubility Equilibrium | $K_{sp}$ |
|---|---|---|
| | **Phosphates** | |
| $Mg_3(PO_4)_2$ | $Mg_3(PO_4)_2(s) \rightleftharpoons 3Mg^{2+}(aq) + 2PO_4^{3-}(aq)$ | $6.3 \times 10^{-26}$ |
| $SrHPO_4$ | $SrHPO_4(s) \rightleftharpoons Sr^{2+}(aq) + HPO_4^{2-}(aq)$ | $1.2 \times 10^{-7}$ |
| $BaHPO_4$ | $BaHPO_4(s) \rightleftharpoons Ba^{2+}(aq) + HPO_4^{2-}(aq)$ | $4.0 \times 10^{-8}$ |
| $LaPO_4$ | $LaPO_4(s) \rightleftharpoons La^{3+}(aq) + PO_4^{3-}(aq)$ | $3.7 \times 10^{-23}$ |
| $Fe_3(PO_4)_2$ | $Fe_3(PO_4)_2(s) \rightleftharpoons 3Fe^{2+}(aq) + 2PO_4^{3-}(aq)$ | $1 \times 10^{-36}$ |
| $Ag_3PO_4$ | $Ag_3PO_4(s) \rightleftharpoons 3Ag^+(aq) + PO_4^{3-}(aq)$ | $2.8 \times 10^{-18}$ |
| $FePO_4$ | $FePO_4(s) \rightleftharpoons Fe^{3+}(aq) + PO_4^{3-}(aq)$ | $4.0 \times 10^{-27}$ |
| $Zn_3(PO_4)_2$ | $Zn_3(PO_4)_2(s) \rightleftharpoons 3Zn^{2+}(aq) + 2PO_4^{3-}(aq)$ | $5 \times 10^{-36}$ |
| $Pb_3(PO_4)_2$ | $Pb_3(PO_4)_2(s) \rightleftharpoons 3Pb^{2+}(aq) + 2PO_4^{3-}(aq)$ | $3.0 \times 10^{-44}$ |
| $Ba_3(PO_4)_2$ | $Ba_3(PO_4)_2(s) \rightleftharpoons 3Ba^{2+}(aq) + 2PO_4^{3-}(aq)$ | $5.8 \times 10^{-38}$ |
| | **Ferrocyanides** | |
| $Zn_2[Fe(CN)_6]$ | $Zn_2[Fe(CN)_6](s) \rightleftharpoons 2Zn^{2+}(aq) + Fe(CN)_6^{4-}(aq)$ | $2.1 \times 10^{-16}$ |
| $Cd_2[Fe(CN)_6]$ | $Cd_2[Fe(CN)_6](s) \rightleftharpoons 2Cd^{2+}(aq) + Fe(CN)_6^{4-}(aq)$ | $4.2 \times 10^{-18}$ |
| $Pb_2[Fe(CN)_6]$ | $Pb_2[Fe(CN)_6](s) \rightleftharpoons 2Pb^{2+}(aq) + Fe(CN)_6^{4-}(aq)$ | $9.5 \times 10^{-19}$ |

## 4.16   COMPLEX ION FORMATION

A **complex ion** consists most commonly of a metal atom or ion, such as $Co^{3+}$, and attached groups called *ligands*. Ligands may be neutral molecules, such as $NH_3$, or anions, such as $Cl^-$. Furthermore, the ligands may be of all one type, as in $[Co(NH_3)_6]^{3+}$, or of different types, as in $[CoCl(NH_3)_5]^{2+}$.

The region surrounding the central atom or ion and its ligands is called the *coordination sphere*. The coordination number is the total number of points at which the central atom or ion attaches to its ligands. Both $[Co(NH_3)_6]^{3+}$ and $[CoCl(NH_3)_5]^{2+}$ have coordination numbers of 6. The most common coordination numbers observed in complex ions are 2, 4, and 6. If the complex carries a net electric charge, as the two examples given here do, it is called a **complex ion**. If it is electrically neutral, it is referred to as a coordination compound. An example of a coordination compound is $[Co(NH_3)_6]Cl_3$.

Many complex ions are colored because the energy differences in the $d$ orbitals match the components of visible light. Substituting one ligand for another produces subtle changes in the energy levels of the $d$ orbitals and striking changes in the colors of complex ions. For example, a solution of $[Cr(H_2O)_6]^{3+}$ is violet while a solution of $[Cr(NH_3)_6]^{3+}$ is yellow.

## 4.17   CHELATING AND SEQUESTERING AGENTS

EDTA is short for ethylenediaminetetraacetic acid. When it acts as a ligand in water, it exists as the ethylenediaminetetraacetate anion $[EDTA]^{4-}$. This anion has a remarkable ability to form complex ions with metal cations in water, which are significantly

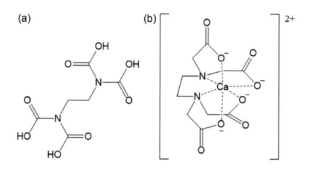

**FIGURE 4.1** (a) The molecular structure of EDTA and (b) EDTA complex of calcium ion.

more stable than ordinary complex ions. It is referred to as a **chelating agent**, and its complex ions are called chelates.

$[EDTA]^{4-}$ will form chelates with hard water ions such as $Fe^{2+}$, $Mg^{2+}$, and $Ca^{2+}$. See Figure 4.1. When it is added to boiler water, it removes these hard water metal ions so effectively that it is referred to as a **sequestering agent**. It actually stops boiler scale buildup. Chelation with $[EDTA]^{4-}$ also prevents the growth of certain bacteria in liquid soaps and shampoos. By sequestering $Ca^{2+}$ and $Mg^{2+}$ ions, which are important constituents of bacterial cell walls, the cell walls disintegrate and the bacteria die.

EDTA is also administered in chelation therapy to remove heavy metals, such as lead or cadmium, from humans who have been exposed to these toxins.

# 5 Chemical Thermodynamics and Thermochemistry

## 5.1 BASIC CONCEPT

**Thermodynamics** is the study of energy transformations. **Thermochemistry** is the study of the thermal energy changes that accompany chemical and physical changes. **Heat** is one form of energy; sound and light are two others that may accompany a chemical reaction. Heat is the energy transferred between objects or systems, which is measured by temperature changes. The total energy is a property of a substance or system; it cannot be measured directly.

Any process occurring within a system that produces or releases heat into the surroundings is called **exothermic**, and its value is given a negative sign by convention. Any process that absorbs heat from the surroundings into the system is called **endothermic**, and its value has a positive sign. Note that an exothermic reaction is not necessarily a spontaneous reaction, and its negative sign should *not* be interpreted as such. This type of interpretation and criterion are reserved for $\Delta G$, the **Gibbs free energy change**, defined in Equation 5.11 and discussed in Equations 5.12 and 5.13.

## 5.2 ENTHALPY AND HESS'S LAW

Since chemical reactions involve the breaking and making of bonds, energy changes generally accompany reactions. To initiate a reaction, an activation energy, $E_a$, must first be supplied for the reactants to reach an activated complex, whether or not the reaction is spontaneous. When a reaction occurs, a change in **enthalpy**, that is, heat content, also known as the **heat of reaction**, $\Delta H_{RX}$, occurs. This is the amount of heat exchanged—either liberated or absorbed—with the surroundings under constant external pressure. It is very helpful to recall Figure 2.1 and a discussion of exothermic and endothermic reactions in Chapter 2, Section 2.7 for clarification. It may be measured experimentally (directly) or calculated theoretically (indirectly) and expressed in either kilojoules (kJ) or kilocalories (kcal).

The enthalpy change known as the **standard heat of formation**, $\Delta H_f^\circ$, is defined as the heat of reaction for a product derived only from its elements, the superscript $^\circ$ meaning measured under standard conditions (usually at STP). See Table 5.1 for thermodynamic values of common substances.

Regardless of the reaction, **Hess's Law** states: The enthalpy change of a chemical reaction is the same regardless of the chemical path taken from reactants to products.

DOI: 10.1201/9781003092759-5

**TABLE 5.1**
**Thermodynamic Constants for the Changes of Formation in Enthalpy,**
**Entropy, and Gibbs Free Energy at STP**

| Aluminum | $\Delta H^\circ_f$ (kJ/mol) | $\Delta G^\circ_f$ (kJ/mol) | $S^\circ$ (J/mol K) |
|---|---|---|---|
| Al (s) | 0 | 0 | 28.3 |
| $AlCl_3$ (s) | −704.2 | −628.8 | 110.7 |
| $Al_2O_3$ (s) | −1675.7 | −1582.3 | 50.9 |
| $Al(OH)_3$ (s) | −1277.0 | | |

| Barium | $\Delta H^\circ_f$ (kJ/mol) | $\Delta G^\circ_f$ (kJ/mol) | $S^\circ$ (J/mol K) |
|---|---|---|---|
| Ba (s) | 0 | 0 | 67.0 |
| $BaCl_2$ (s) | −858.6 | −810.4 | 123.7 |
| $BaCO_3$ (s) | −1219.0 | −1139.0 | 112.0 |
| BaO (s) | −553.5 | −525.1 | 70.4 |
| $Ba(OH)_2$ (s) | −946.0 | | |
| $BaSO_4$ (s) | −1473.2 | −1362.2 | 132.2 |

| Beryllium | $\Delta H^\circ_f$ (kJ/mol) | $\Delta G^\circ_f$ (kJ/mol) | $S^\circ$ (J/mol K) |
|---|---|---|---|
| Be (s) | 0 | 0 | 9.5 |
| BeO (s) | −599.0 | −569.0 | 14.0 |
| $Be(OH)_2$ (s) | −902.5 | −815.0 | 51.9 |

| Bromine | $\Delta H^\circ_f$ (kJ/mol) | $\Delta G^\circ_f$ (kJ/mol) | $S^\circ$ (J/mol K) |
|---|---|---|---|
| Br (g) | 111.9 | 82.4 | 175.0 |
| $Br_2$ (l) | 0 | 0 | 152.2 |
| $Br_2$ (g) | 30.9 | 3.1 | 245.5 |
| $Br_2$ (aq) | −3.0 | 4.0 | 130.0 |
| Br− (aq) | −121.0 | −175.0 | 82.0 |
| $BrF_3$ (g) | −255.6 | −229.4 | 292.5 |
| HBr (g) | −36.3 | −53.5 | 198.7 |

| Cadmium | $\Delta H^\circ_f$ (kJ/mol) | $\Delta G^\circ_f$ (kJ/mol) | $S^\circ$ (J/mol K) |
|---|---|---|---|
| Cd (s) | 0 | 0 | 52.0 |
| CdO (s) | −258.0 | −228.0 | 55.0 |
| $Cd(OH)_2$ (s) | −561.0 | −474.0 | 96.0 |
| CdS (s) | −162.0 | −156.0 | 65.0 |
| $CdSO_4$ (s) | −935.0 | −823.0 | 123.0 |

| Calcium | $\Delta H^\circ_f$ (kJ/mol) | $\Delta G^\circ_f$ (kJ/mol) | $S^\circ$ (J/mol K) |
|---|---|---|---|
| Ca (s) | 0 | 0 | 41.4 |
| Ca (g) | 178.2 | 144.3 | 158.9 |
| $Ca^{2+}$ (g) | 1925.9 | | |
| $CaC_2$ (s) | −59.8 | −64.9 | 70.0 |
| $CaCO_3$ (s, calcite) | −1206.9 | −1128.8 | 92.9 |
| $CaCl_2$ (s) | −795.8 | −748.1 | 104.6 |
| $CaF_2$ (s) | −1219.6 | −1167.3 | 68.9 |
| $CaH_2$ (s) | −186.2 | −147.2 | 42.0 |

(*Continued*)

**TABLE 5.1 (*Continued*)**

**Thermodynamic Constants for the Changes of Formation in Enthalpy, Entropy, and Gibbs Free Energy at STP**

| | | | |
|---|---|---|---|
| CaO (s) | −635.1 | −604.0 | 39.8 |
| CaS (s) | −482.4 | −477.4 | 56.5 |
| Ca(OH)$_2$ (s) | −986.1 | −898.5 | 83.4 |
| Ca(OH)$_2$ (aq) | −1002.8 | −868.1 | −74.5 |
| Ca$_3$(PO$_4$)$_2$ (s) | −4126.0 | −3890.0 | 241.0 |
| CaSO$_4$ (s) | −1434.1 | −1321.8 | 106.7 |
| CaSiO$_3$ (s) | −1630.0 | −1550.0 | 84.0 |
| **Carbon** | **ΔH°$_f$ (kJ/mol)** | **ΔG°$_f$ (kJ/mol)** | **S° (J/mol K)** |
| C (s, graphite) | 0 | 0 | 5.7 |
| C (s, diamond) | 1.9 | 2.9 | 2.4 |
| C (g) | 716.7 | 671.3 | 158.1 |
| CCl$_4$ (l) | −135.4 | −65.2 | 216.4 |
| CCl$_4$ (g) | −102.9 | −60.6 | 309.9 |
| CHCl$_3$ (l) | −134.5 | −73.7 | 201.7 |
| CHCl$_3$ (g) | −103.1 | −70.3 | 295.7 |
| CH$_4$ (g) | −74.8 | −50.7 | 186.3 |
| CH$_3$OH (g) | −200.7 | −162.0 | 239.8 |
| CH$_3$OH (l) | −238.7 | −166.3 | 126.8 |
| H$_2$CO (g) | −116.0 | −110.0 | 219.0 |
| HCOOH (g) | −363.0 | −351.0 | 249.0 |
| HCN (g) | 135.1 | 125.0 | 202.0 |
| C$_2$H$_2$ (g) | 226.7 | 209.2 | 200.9 |
| C$_2$H$_4$ (g) | 52.3 | 68.2 | 219.6 |
| CH$_3$CHO (g, acetaldehyde) | −166.0 | −129.0 | 250.0 |
| C$_2$H$_4$O (g, ethylene oxide) | −53.0 | −13.0 | 242.0 |
| CH$_3$CH$_2$OH (l) | −277.7 | −174.8 | 160.7 |
| CH$_3$CH$_2$OH (g) | −235.1 | −168.5 | 282.7 |
| CH$_3$COOH (l) | −484.0 | −389.0 | 160.0 |
| C$_2$H$_6$ (g) | −84.7 | −32.8 | 229.6 |
| C$_3$H$_6$ (g) | 20.9 | 62.7 | 266.9 |
| C$_3$H$_8$ (g) | −103.8 | −23.5 | 269.9 |
| CH$_2$=CHCN (l) | 152.0 | 190.0 | 274.0 |
| C$_6$H$_6$ (l) | 49.0 | 124.5 | 172.8 |
| C$_6$H$_{12}$O$_6$ (s) | −1275.0 | −911.0 | 212.0 |
| CO (g) | −110.5 | −137.2 | 197.7 |
| CO$_2$ (g) | −393.5 | −394.4 | 213.7 |
| CS$_2$ (g) | 117.4 | 67.1 | 237.8 |
| COCl$_2$ (g) | −218.8 | −204.6 | 283.5 |

(*Continued*)

**TABLE 5.1 (*Continued*)**

**Thermodynamic Constants for the Changes of Formation in Enthalpy, Entropy, and Gibbs Free Energy at STP**

| Chlorine | $\Delta H^\circ_f$ (kJ/mol) | $\Delta G^\circ_f$ (kJ/mol) | $S^\circ$ (J/mol K) |
|---|---|---|---|
| Cl (g) | 121.7 | 105.7 | 165.2 |
| $Cl_2$ (g) | 0 | 0 | 223.1 |
| $Cl_2$ (aq) | −23.0 | 7.0 | 121.0 |
| $Cl^-$ (aq) | −167.0 | −131.0 | 57.0 |
| $Cl^-$ (g) | −233.1 | | |
| HCl (g) | −92.3 | −95.3 | 186.9 |
| HCl (aq) | −167.2 | −131.2 | 56.5 |

| Chromium | $\Delta H^\circ_f$ (kJ/mol) | $\Delta G^\circ_f$ (kJ/mol) | $S^\circ$ (J/mol K) |
|---|---|---|---|
| Cr (s) | 0 | 0 | 23.8 |
| $Cr_2O_3$ (s) | −1139.7 | −1058.1 | 81.2 |
| $CrO_3$ (s) | −579.0 | −502.0 | 72.0 |
| $CrCl_3$ (s) | −556.5 | −486.1 | 123.0 |

| Copper | $\Delta H^\circ_f$ (kJ/mol) | $\Delta G^\circ_f$ (kJ/mol) | $S^\circ$ (J/mol K) |
|---|---|---|---|
| Cu (s) | 0 | 0 | 33.2 |
| $CuCl_2$ (s) | −220.1 | −175.7 | 108.1 |
| $CuCO_3$ (s) | −595.0 | −518.0 | 88.0 |
| $Cu_2O$ (s) | −170.0 | −148.0 | 93.0 |
| CuO (s) | −157.3 | −129.7 | 42.6 |
| $Cu(OH)_2$ (s) | −450.0 | −372.0 | 108.0 |
| CuS (s) | −49.0 | −49.0 | 67.0 |

| Fluorine | $\Delta H^\circ_f$ (kJ/mol) | $\Delta G^\circ_f$ (kJ/mol) | $S^\circ$ (J/mol K) |
|---|---|---|---|
| $F_2$ (g) | 0 | 0 | 202.8 |
| F (g) | 79.0 | 61.9 | 158.8 |
| $F^-$ (g) | −255.4 | | |
| $F^-$ (aq) | −332.6 | −278.8 | −13.8 |
| HF (g) | −271.1 | −273.2 | 173.8 |
| HF (aq) | −332.6 | −278.8 | 88.7 |

| Hydrogen | $\Delta H^\circ_f$ (kJ/mol) | $\Delta G^\circ_f$ (kJ/mol) | $S^\circ$ (J/mol K) |
|---|---|---|---|
| $H_2$ (g) | 0 | 0 | 130.7 |
| H (g) | 218.0 | 203.2 | 114.7 |
| $H^+$ (g) | 1536.2 | | |
| $H^+$ (aq) | 0 | | |
| $OH^-$ (aq) | −230.0 | −157.0 | −11.0 |
| $H_2O$ (l) | −285.8 | −237.1 | 69.9 |
| $H_2O$ (g) | −241.8 | −228.6 | 188.8 |
| $H_2O_2$ (l) | −187.8 | −120.4 | 109.6 |

(*Continued*)

## TABLE 5.1 (*Continued*)
## Thermodynamic Constants for the Changes of Formation in Enthalpy, Entropy, and Gibbs Free Energy at STP

| Iodine | $\Delta H^{\circ}_f$ (kJ/mol) | $\Delta G^{\circ}_f$ (kJ/mol) | $S^{\circ}$ (J/mol K) |
|---|---|---|---|
| $I_2$ (s) | 0 | 0 | 116.1 |
| $I_2$ (g) | 62.4 | 19.3 | 260.7 |
| $I_2$ (aq) | 23.0 | 16.0 | 137 |
| I (g) | 106.8 | 70.3 | 180.8 |
| $I^-$ (g) | −197.0 | | |
| $I^-$ (aq) | −55.0 | −52.0 | 106.0 |
| ICl (g) | 17.8 | −5.5 | 247.6 |

| Iron | $\Delta H^{\circ}_f$ (kJ/mol) | $\Delta G^{\circ}_f$ (kJ/mol) | $S^{\circ}$ (J/mol K) |
|---|---|---|---|
| Fe (s) | 0 | 0 | 27.8 |
| $Fe_3C$ (s) | 21.0 | 15.0 | 108.0 |
| $FeCl_2$ (s) | −341.8 | −302.3 | 118.0 |
| $FeCl_3$ (s) | −399.5 | −333.9 | 142.3 |
| $Fe_{0.95}O$ (s) (wustite) | −264.0 | −240.0 | 59.0 |
| FeO (s) | −272.0 | | |
| $Fe_3O_4$ (s, magnetite) | −1118.4 | −1015.4 | 146.4 |
| $Fe_2O_3$ (s, hematite) | −824.2 | −742.2 | 87.4 |
| FeS (s) | −95.0 | −97.0 | 67.0 |
| $FeS_2$ (s, pyrite) | −178.2 | −166.9 | 52.9 |
| $FeSO_4$ (s) | −929.0 | −825.0 | 121.0 |
| $Fe(CO)_5$ (l) | −774.0 | −705.3 | 338.1 |

| Lead | $\Delta H^{\circ}_f$ (kJ/mol) | $\Delta G^{\circ}_f$ (kJ/mol) | $S^{\circ}$ (J/mol K) |
|---|---|---|---|
| Pb (s) | 0 | 0 | 64.8 |
| $PbCl_2$ (s) | −359.4 | −314.1 | 136.0 |
| PbO (s, yellow) | −217.3 | −187.9 | 68.7 |
| $PbO_2$ (s) | −277.0 | −217.0 | 69.0 |
| PbS (s) | −100.4 | −98.7 | 91.2 |
| $PbSO_4$ (s) | −920.0 | −813.0 | 149.0 |

| Magnesium | $\Delta H^{\circ}_f$ (kJ/mol) | $\Delta G^{\circ}_f$ (kJ/mol) | $S^{\circ}$ (J/mol K) |
|---|---|---|---|
| Mg (s) | 0 | 0 | 32.7 |
| $MgCl_2$ (s) | −641.3 | −591.8 | 89.6 |
| $MgCO_3$ (s) | −1095.8 | −1012.1 | 65.7 |
| MgO (s) | −601.7 | −569.4 | 26.9 |
| $Mg(OH)_2$ (s) | −924.5 | −833.5 | 63.2 |
| MgS (s) | −346.0 | −341.8 | 50.3 |

| Manganese | $\Delta H^{\circ}_f$ (kJ/mol) | $\Delta G^{\circ}_f$ (kJ/mol) | $S^{\circ}$ (J/mol K) |
|---|---|---|---|
| Mn (s) | 0 | 0 | 32.0 |
| MnO (s) | −385.0 | −363.0 | 60.0 |
| $Mn_3O_4$ (s) | −1387.0 | −1280.0 | 149.0 |

*(Continued)*

**TABLE 5.1 (*Continued*)**

**Thermodynamic Constants for the Changes of Formation in Enthalpy, Entropy, and Gibbs Free Energy at STP**

| | $\Delta H^\circ_f$ (kJ/mol) | $\Delta G^\circ_f$ (kJ/mol) | $S^\circ$ (J/mol K) |
|---|---|---|---|
| $Mn_2O_3$ (s) | −971.0 | −893.0 | 110.0 |
| $MnO_2$ (s) | −521.0 | −466.0 | 53.0 |
| $MnO_4^-$ (aq) | −543.0 | −449.0 | 190.0 |
| **Mercury** | $\Delta H^\circ_f$ (kJ/mol) | $\Delta G^\circ_f$ (kJ/mol) | $S^\circ$ (J/mol K) |
| Hg (l) | 0 | 0 | 75.9 |
| $HgCl_2$ (s) | −224.3 | −178.6 | 146.0 |
| $Hg_2Cl_2$ (s) | −265.4 | −210.7 | 191.7 |
| HgO (s, red) | −90.8 | −58.5 | 70.3 |
| HgS (s, red) | −58.2 | −50.6 | 82.4 |
| **Nickel** | $\Delta H^\circ_f$ (kJ/mol) | $\Delta G^\circ_f$ (kJ/mol) | $S^\circ$ (J/mol K) |
| Ni (s) | 0 | 0 | 29.9 |
| $NiCl_2$ (s) | −305.3 | −259.0 | 97.7 |
| NiO (s) | −239.7 | −211.7 | 38.0 |
| $Ni(OH)_2$ (s) | −538.0 | −453.0 | 79.0 |
| NiS (s) | −93.0 | −90.0 | 53.0 |
| **Nitrogen** | $\Delta H^\circ_f$ (kJ/mol) | $\Delta G^\circ_f$ (kJ/mol) | $S^\circ$ (J/mol K) |
| $N_2$ (g) | 0 | 0 | 191.6 |
| N (g) | 472.7 | 455.6 | 153.3 |
| $NH_3$ (g) | −46.1 | −16.5 | 192.5 |
| $NH_3$ (aq) | −80.0 | −27.0 | 111.0 |
| $NH_4^+$ (aq) | −132.0 | −79.0 | 113.0 |
| NO (g) | 90.3 | 86.6 | 210.8 |
| NOCl (g) | 51.7 | 66.1 | 261.8 |
| $NO_2$ (g) | 33.9 | 51.3 | 240.1 |
| $N_2O$ (g) | 82.1 | 104.2 | 219.9 |
| $N_2O_4$ (g) | 9.2 | 97.9 | 304.3 |
| $N_2O_4$ (l) | −19.5 | 97.0 | 209.0 |
| $N_2O_5$ (s) | −42.0 | 134.0 | 178.0 |
| $N_2H_4$ (l) | 50.6 | 149.3 | 121.2 |
| $N_2H_3CH_3$ (l) | 54.0 | 180.0 | 166.0 |
| $HNO_3$ (aq) | −207.4 | −111.3 | 146.4 |
| $HNO_3$ (l) | −174.1 | −80.7 | 155.6 |
| $HNO_3$ (g) | −135.1 | −74.7 | 266.4 |
| $NH_4ClO_4$ (s) | −295.0 | −89.0 | 186.0 |
| $NH_4Cl$ (s) | −314.4 | −202.9 | 94.6 |
| $NH_4Cl$ (aq) | −299.7 | −210.5 | 169.9 |
| $NH_4NO_3$ (s) | −365.6 | −183.9 | 151.1 |
| $NH_4NO_3$ (aq) | −339.9 | −190.6 | 259.8 |

(*Continued*)

**TABLE 5.1 (*Continued*)**

**Thermodynamic Constants for the Changes of Formation in Enthalpy, Entropy, and Gibbs Free Energy at STP**

| Oxygen | $\Delta H^\circ_f$ (kJ/mol) | $\Delta G^\circ_f$ (kJ/mol) | $S^\circ$ (J/mol K) |
|---|---|---|---|
| $O_2$ (g) | 0 | 0 | 205.1 |
| O (g) | 249.2 | 231.7 | 161.1 |
| $O_3$ (g) | 142.7 | 163.2 | 238.9 |

| Phosphorus | $\Delta H^\circ_f$ (kJ/mol) | $\Delta G^\circ_f$ (kJ/mol) | $S^\circ$ (J/mol K) |
|---|---|---|---|
| P (s, white) | 0 | 0 | 164.4 |
| P (s, red) | −70.4 | −48.4 | 91.2 |
| P (s, black) | −39.0 | −33.0 | 23.0 |
| P (g) | 314.6 | 278.3 | 163.2 |
| $P_4$ (s, white) | 0 | 0 | 41.1 |
| $P_4$ (s, red) | −17.6 | −12.1 | 22.8 |
| $P_4$ (g) | 59.0 | 24.0 | 280.0 |
| $PF_5$ (g) | −1578.0 | −1509.0 | 296.0 |
| $PH_3$ (g) | 5.4 | 13.4 | 210.2 |
| $PCl_3$ (g) | −287.0 | −267.8 | 311.8 |
| $H_3PO_4$ (l) | −1279.0 | −1119.1 | 110.5 |
| $H_3PO_4$ (aq) | −1288.0 | −1143.0 | 158.0 |
| $P_4O_{10}$ (s) | −2984.0 | −2697.7 | 228.9 |

| Potassium | $\Delta H^\circ_f$ (kJ/mol) | $\Delta G^\circ_f$ (kJ/mol) | $S^\circ$ (J/mol K) |
|---|---|---|---|
| K (s) | 0 | 0 | 64.2 |
| KCl (s) | −436.7 | −409.1 | 82.6 |
| $KClO_3$ (s) | −397.7 | −296.3 | 143.1 |
| $KClO_4$ (s) | −433.0 | −304.0 | 151.0 |
| KI (s) | −327.9 | −324.9 | 106.3 |
| $K_2O$ (s) | −361.0 | −322.0 | 98.0 |
| $K_2O_2$ (s) | −496.0 | −430.0 | 113.0 |
| $KO_2$ (s) | −283.0 | −238.0 | 117.0 |
| KOH (s) | −424.8 | −379.1 | 78.9 |
| KOH (aq) | −482.4 | −440.5 | 91.6 |

| Silicon | $\Delta H^\circ_f$ (kJ/mol) | $\Delta G^\circ_f$ (kJ/mol) | $S^\circ$ (J/mol K) |
|---|---|---|---|
| Si (s) | 0 | 0 | 18.3 |
| $SiBr_4$ (l) | −457.3 | −443.9 | 277.8 |
| SiC (s) | −65.3 | −62.8 | 16.6 |
| $SiCl_4$ (g) | −657.0 | −617.0 | 330.7 |
| $SiH_4$ (g) | 34.3 | 56.9 | 204.6 |
| $SiF_4$ (g) | −1614.9 | −1572.7 | 282.5 |
| $SiO_2$ (s, quartz) | −910.9 | −856.6 | 41.8 |

| Silver | $\Delta H^\circ_f$ (kJ/mol) | $\Delta G^\circ_f$ (kJ/mol) | $S^\circ$ (J/mol K) |
|---|---|---|---|
| Ag (s) | 0 | 0 | 42.6 |
| $Ag^+$ (aq) | 105.0 | 77.0 | 73.0 |

*(Continued)*

**TABLE 5.1 (*Continued*)**
**Thermodynamic Constants for the Changes of Formation in Enthalpy, Entropy, and Gibbs Free Energy at STP**

| | $\Delta H^\circ_f$ (kJ/mol) | $\Delta G^\circ_f$ (kJ/mol) | $S^\circ$ (J/mol K) |
|---|---|---|---|
| AgBr (s) | −100.0 | −97.0 | 107.0 |
| AgCN (s) | 146.0 | 164.0 | 84.0 |
| AgCl (s) | −127.1 | −109.8 | 96.2 |
| $Ag_2CrO_4$ (s) | −712.0 | −622.0 | 217.0 |
| AgI (s) | −62.0 | −66.0 | 115.0 |
| $Ag_2O$ (s) | −31.1 | −11.2 | 121.3 |
| $AgNO_3$ (s) | −124.4 | −33.4 | 140.9 |
| $Ag_2S$ (s) | −32.0 | −40.0 | 146.0 |
| **Sodium** | $\Delta H^\circ_f$ (kJ/mol) | $\Delta G^\circ_f$ (kJ/mol) | $S^\circ$ (J/mol K) |
| Na (s) | 0 | 0 | 51.2 |
| Na (g) | 107.3 | 76.8 | 153.7 |
| $Na^+$ (g) | 609.4 | | |
| $Na^+$ (aq) | −240.0 | −262.0 | 59.0 |
| NaBr (s) | −361.0 | −349.0 | 86.8 |
| $Na_2CO_3$ (s) | −1130.7 | −1044.4 | 135.0 |
| $NaHCO_3$ (s) | −948.0 | −852.0 | 102.0 |
| NaCl (s) | −411.2 | −384.1 | 72.1 |
| NaCl (g) | −176.7 | −196.7 | 229.8 |
| NaCl (aq) | −407.3 | −393.1 | 115.5 |
| NaH (s) | −56.0 | −33.0 | 40.0 |
| NaI (s) | −288.0 | −282.0 | 91.0 |
| $NaNO_2$ (s) | −359.0 | | |
| $NaNO_3$ (s) | −467.0 | −366.0 | 116.0 |
| $Na_2O$ (s) | −416.0 | −377.0 | 73.0 |
| $Na_2O_2$ (s) | −515.0 | −451.0 | 95.0 |
| NaOH (s) | −425.6 | −379.5 | 64.5 |
| NaOH (aq) | −470.1 | −419.2 | 48.1 |
| **Sulfur** | $\Delta H^\circ_f$ (kJ/mol) | $\Delta G^\circ_f$ (kJ/mol) | $S^\circ$ (J/mol K) |
| S (s, rhombic) | 0 | 0 | 31.8 |
| S (s, monoclinic) | 0.3 | 0.1 | 33.0 |
| S (g) | 278.8 | 238.3 | 167.8 |
| $S_2^-$ (aq) | 33.0 | 86.0 | −15.0 |
| $S_8$ (g) | 102.0 | 50.0 | 431.0 |
| $S_2Cl_2$ (g) | −18.4 | −31.8 | 331.5 |
| $SF_6$ (g) | −1209.0 | −1105.3 | 291.8 |
| $H_2S$ (g) | −20.6 | −33.6 | 205.8 |
| $SO_2$ (g) | −296.8 | −300.2 | 248.2 |
| $SO_3$ (g) | −395.7 | −371.1 | 256.8 |
| $SOCl_2$ (g) | −212.5 | −198.3 | 309.8 |
| $SO_4^{2-}$ (aq) | −909.0 | −745.0 | 20.0 |

(*Continued*)

**TABLE 5.1 (*Continued*)**
**Thermodynamic Constants for the Changes of Formation in Enthalpy,**
**Entropy, and Gibbs Free Energy at STP**

| | $\Delta H^\circ_f$ (kJ/mol) | $\Delta G^\circ_f$ (kJ/mol) | $S^\circ$ (J/mol K) |
|---|---|---|---|
| $H_2SO_4$ (l) | −814.0 | −690.0 | 156.9 |
| $H_2SO_4$ (aq) | −909.3 | −744.5 | 20.1 |
| **Tin** | **$\Delta H^\circ_f$ (kJ/mol)** | **$\Delta G^\circ_f$ (kJ/mol)** | **$S^\circ$ (J/mol K)** |
| Sn (s, white) | 0 | 0 | 51.6 |
| Sn (s, gray) | −2.1 | 0.1 | 44.1 |
| $SnCl_4$ (l) | −511.3 | −440.1 | 258.6 |
| $SnCl_4$ (g) | −471.5 | −432.2 | 365.8 |
| SnO (s) | −285.0 | −257.0 | 56.0 |
| $SnO_2$ (s) | −580.7 | −519.6 | 52.3 |
| $Sn(OH)_2$ (s) | −561.0 | −492.0 | 155.0 |
| **Titanium** | **$\Delta H^\circ_f$ (kJ/mol)** | **$\Delta G^\circ_f$ (kJ/mol)** | **$S^\circ$ (J/mol K)** |
| Ti (s) | 0 | 0 | 30.6 |
| $TiCl_4$ (l) | −804.2 | −737.2 | 252.3 |
| $TiCl_4$ (g) | −763.2 | −726.7 | 354.9 |
| $TiO_2$ (s) | −939.7 | −884.5 | 49.9 |
| **Uranium** | **$\Delta H^\circ_f$ (kJ/mol)** | **$\Delta G^\circ_f$ (kJ/mol)** | **$S^\circ$ (J/mol K)** |
| U (s) | 0 | 0 | 50.0 |
| $UF_6$ (s) | −2137.0 | −2008.0 | 228.0 |
| $UF_6$ (g) | −2113.0 | −2029.0 | 380.0 |
| $UO_2$ (s) | −1084.0 | −1029.0 | 78.0 |
| $U_3O_8$ (s) | −3575.0 | −3393.0 | 282.0 |
| $UO_3$ (s) | −1230.0 | −1150.0 | 99.0 |
| **Xenon** | **$\Delta H^\circ_f$ (kJ/mol)** | **$\Delta G^\circ_f$ (kJ/mol)** | **$S^\circ$ (J/mol K)** |
| Xe (g) | 0 | 0 | 170.0 |
| $XeF_2$ (g) | −108.0 | −48.0 | 254.0 |
| $XeF_4$ (s) | −251.0 | −121.0 | 146.0 |
| $XeF_6$ (g) | −294.0 | | |
| $XeO_3$ (s) | 402.0 | | |
| **Zinc** | **$\Delta H^\circ_f$ (kJ/mol)** | **$\Delta G^\circ_f$ (kJ/mol)** | **$S^\circ$ (J/mol K)** |
| Zn (s) | 0 | 0 | 41.6 |
| $ZnCl_2$ (s) | −415.1 | −369.4 | 111.5 |
| ZnO (s) | −348.3 | −318.3 | 43.6 |
| $Zn(OH)_2$ (s) | −642.0 | | |
| ZnS (s, wurtzite) | −193.0 | | |
| ZnS (s, zinc blende) | −206.0 | −201.3 | 57.7 |
| $ZnSO_4$ (s) | −983.0 | −874.0 | 120.0 |

If the reaction of interest is the sum of several other reactions, the individual heats of reaction must be added algebraically. In other words,

$$\Delta H_{RX} = \sum \Delta H_{products} - \sum \Delta H_{reactants} \qquad (5.1)$$

Note that this same principle also applies to the calculation of the Gibbs free energy change for chemical reaction.

**Example 5.1:** Using Hess' Law To Compute Enthalpy Change

The standard heats of formation $\Delta H_f^o$, for $NO_2$ and $N_2O_4$, are as follows: $\Delta H_f^o$ [$NO_2$] = 33.9 kJ/mole, $\Delta H_f^o$ [$N_2O_4$] = −19.5 kJ/mole. Predict by calculation the heat of reaction, $\Delta H_{RX}$, for the following reaction at standard conditions.

$$2NO_2(g) \rightleftharpoons N_2O_4(g)$$

*Solution*

Use Hess's Law.

$$\Delta H_{RX^o} = \sum \Delta H_{products} - \sum \Delta H_{reactants}$$

$$= \Delta H_{f^o}[N_2O_4] - 2\Delta H_{f^o}[NO_2]$$

$$= (1.00\ mol)(-19.5\ kJ/mol) - (2.00\ mol)(33.9\ kJ/mol)$$

$$= -87.3\ kJ$$

Note that the negative sign means that the reaction is exothermic.

**Example 5.2:** Using Hess' Law To Compute Enthalpy Change

Given the following reactions and data for $\Delta H^o$:

$$N_2O_4(g) \rightarrow 2NO_2 \Delta H^o(I) = 87.30\ kJ$$

$$2NO(g) + O_2(g) \rightarrow 2NO_2 \Delta H^o(II) = -114.14\ kJ$$

Compute $\Delta H_{RX}^o$ for the following reaction:

$$2NO(g) + O_2(g) \rightarrow N_2O_4(g)$$

*Solution*

First, recognize that, if the first given reaction is reversed (I) and added to the second reaction (II), the desired reaction is obtained, because the $2NO_2$ terms will be on opposite sides of the reaction and will thus cancel. To do this, the sign of

$\Delta H^o$(I) must be changed from positive to negative, so the value becomes $-87.30$ kJ. Then, add the two $\Delta H^o$ values together. Thus:

$$\Delta H_{RX^o} = \Delta H^o\left(I\right) + \Delta H^o\left(II\right)$$

$$= -87.30\,\text{kJ} + \left(-114.14\,\text{kJ}\right) = -201.44\,\text{kJ}$$

Another way to solve this problem is to use Hess's Law and set the solution up as follows:

$$\Delta H_{RX^o} = \sum \Delta H_{\text{products}} - \sum \Delta H_{\text{reactants}}$$

$$= \Delta H_{f^o}\left[N_2O_4\right] - \left\{2\Delta H_{f^o}\left[NO\right] + \Delta H_{f^o}\left[O_2\right]\right\}$$

The problem with this approach is that, while $\Delta H_f^o$ $[O_2] = 0$ and the value for $\Delta H_f^o$ $[N_2O_4]$ is given in Example 5.2, the value for $\Delta H_f^o$ $[NO]$ is unknown.

## 5.3   THE FIRST LAW AND THE CONSERVATION OF ENERGY

In understanding the laws of thermodynamics, it is useful to define *system* and distinguish it from *surroundings* very carefully. In general, there are three types of systems:

1. **Closed system**—a system that no mass enters or leaves
2. **Isolated system**—a system that no mass or heat enters or leaves.
3. **Open system**—a system that mass or heat can enter or leave.

Furthermore, there are two kinds of *boundaries/walls* or processes:

4. **Diathermal walls**—walls that permit the transfer of heat between system and surroundings.
5. **Adiabatic walls**—walls that do not permit the transfer of heat between system and surroundings (e.g., good insulators).

The **first law of thermodynamics** can be stated in several ways. For example:

- The energy of the universe is constant. The total of the energy change of the system plus the energy change of the surroundings is, therefore, also constant.
- During a chemical or physical change in any defined system, energy can neither be created nor be destroyed, but only changes form.

Expressed in mathematical terms:

$$\Delta E = q - w \tag{5.2}$$

where
  $\Delta E$ = the total internal energy change of the system,
  $q$ = the heat absorbed *by* the system *from* the surroundings,
  $w$ = the work done *by* the system *on* the surroundings.

For adiabatic processes, $q = 0$, so $\Delta E = -w$.

For isothermal changes in ideal gases, $\Delta E = 0$, since $E$ is a function of temperature only. Therefore, $q = w$.

For pressure–volume (i.e., mechanical) work of an ideal gas (i.e., expansion or contraction) if pressure $P$ is constant:

$$w = \int P \, dV = P\Delta V \qquad (5.3)$$

If pressure P is not constant (i.e., variable), then:

$$w = \int P \, dV = \int \frac{nRT}{V} \, dV$$

$$= nRT \ln \frac{V_2}{V_1} \qquad (5.4)$$

$\Delta E$ and $\Delta H$ are related by the equation $\Delta H = \Delta E + P \, \Delta V$ at constant $P$.

Other types of thermodynamic work may include electrical and magnetic work.

**Example 5.3:** Using the First Law of Thermodynamics (Energy Change)

Ten moles of an ideal gas in a piston and cylinder assembly absorbs 2500 J of heat. The gas expands from 2.0 to 8.5 L against a constant external pressure of 2.5 atm. What is the internal energy change of the system?

*Solution*

$$E = q - w = q - P\Delta V$$

$$= 2500\,\mathrm{J} - \left(2.5\,\mathrm{atm}\right)\left(8.5\,\mathrm{L} - 2.0\,\mathrm{L}\right)$$

$$= 2500\,\mathrm{J} - 16.25\,\mathrm{L*atm} \times \left(101.3\,\mathrm{J/L*atm}\right)$$

$$= 2500\,\mathrm{J} - 1646\,\mathrm{J} = 854\,\mathrm{J}$$

Since the answer is positive, this amount of energy is *gained* by the system (and concurrently *lost* by the surroundings to the system in an equal amount).

## 5.4   THE SECOND LAW AND ENTROPY

The **second law of thermodynamics** may be expressed in at least two different ways. For example:

- The **entropy, S,** that is, the randomness or disorder, of the universe either stays the same or increases, but never decreases.
- During a physical change or chemical reaction, the *total* **entropy change,** $\Delta S_{total}$, of the system plus the surroundings is either zero or positive,

but never negative. The entropy change of a system, $\Delta S_{sys}$, may be calculated as the reversible heat, $q_{rev}$, either gained or lost by a system, divided by the absolute temperature T:

$$\Delta S_{sys} = \frac{q_{rev}}{T} \qquad (5.5)$$

Two consequences of the second law are as follows:

• The construction of a perpetual motion machine is impossible.
• It is impossible to build a heat engine with 100% efficiency.

A corollary of the second law, often referred to as the **zeroth law**, states: Heat always flows from regions of hot to regions of cold *spontaneously*.

The term "reversible" refers to a process that is carried out over an infinitely slow series of steps such that each step is in equilibrium with a previous step. Practically speaking, other words for reversible are *gradual* or *extremely slow*.

As a second component of thermodynamics, entropy is a measure of the statistical disorder or randomness of a system. The universe tends to move toward greater total disorder, and this is expressed in terms of entropy or $S$. Entropy, unlike enthalpy, can be found explicitly. Entropy specifically is a measure of the number of microstates available to a chemical system. Microstates are individual possible states of the system, where a state is a particular arrangement of positions for particles and a particular distribution of kinetic energy among those particles. This can be calculated as shown in Equation 5.6, where $S$ is the entropy, $k$ is Boltzmann's constant, and $W$ is the number of available microstates.

$$S = k \ln W \qquad (5.6)$$

Entropy increases as the size of molecules increases, as the number of molecules increases, as the temperature increases, and as the volume particles occupy increases. Each of these factors increases the number of available microstates. Just as for enthalpy, the change in entropy can be calculated from the entropy inherent in starting materials and products, as shown in Equation 5.7.

$$\Delta S_{reaction} = \left[ \text{sum of } \Delta S_f^{\circ} \text{ of products} \right] - \left[ \text{sum of } \Delta S_f^{\circ} \text{ of reactants} \right] \qquad (5.7)$$

For all real or natural changes, which are spontaneous and irreversible, the entropy of a system increases, so the entropy *change*, $\Delta S$, is positive, never zero or negative. On the grand scale, this means that the total is really the entire universe. Mathematically, this may be expressed as follows:

$$\Delta S_{total} = \Delta S_{system} + \Delta S_{surroundings} = \Delta S_{universe} \geq 0 \qquad (5.8)$$

For phase transitions, $\Delta S_{trans} = \dfrac{q_{trans}}{T} = \dfrac{-\Delta H_{trans}}{T}$ where $\Delta H_{trans}$ may equal $\Delta H_{vap}$ or $\Delta H_{fus}$

For the heating of any substance, without a phase transition, the heat gained or lost, $q$, may be expressed as

$$Q = (m)(C)(\Delta T) = mC\Delta T \tag{5.9}$$

where
   $m$ = the mass of the substance in grams or moles,
   $C$ = the specific heat or heat capacity in suitable units,
   $\Delta T$ = the change in temperature in degrees Celsius or Kelvins.

For other processes involving changes in temperature and/or volume:

$$\Delta S = C_v \ln\frac{T_2}{T_1} + nR \ln\frac{V_2}{V_1} \tag{5.10}$$

**Example 5.4:** Using the Second Law of Thermodynamics (Entropy Change)

One mole of an ideal gas expands reversibly and isothermally from an initial volume of 2 L to a final volume of 20 L. Calculate the entropy change, in calories/Kelvin, of the system and the surroundings.

*Solution*

Since this process is isothermal, $T_1 = T_2$, use $\Delta S = nR \ln V_2 / V_1$. Then:

$$\Delta S_{sys} = (1.00\,\text{mol})(1.99\,\text{cal/K}\cdot\text{mol})\ln\left(\frac{20\,\text{L}}{2.0\,\text{L}}\right)$$

$$= 4.58\,\text{cal/K}$$

Since this is a reversible process:

$$\Delta S_{sys} = -\Delta S_{surr} = -4.58\,\text{cal/k}$$

**Example 5.5:** Using the Second Law of Thermodynamics (Entropy Change)

Calculate the entropy change when 100 g of water vaporizes at 100°C.

*Solution*

This is a phase transition. Use $\Delta S_{trans} = \Delta H_{vap} / T$, (mentioned above for phase transitions) where $\Delta\Delta H_{vap} = 540\,\text{cal/g}$ and $T = 373$ K. Then:

$$\Delta S = \frac{\left(540\dfrac{\text{cal}}{\text{g}}\right)(100\,\text{g})}{373\,\text{K}}$$

$$= 145\,\text{cal/K}$$

## 5.5   THE THIRD LAW AND ABSOLUTE ZERO

The **third law of thermodynamics** provides a basis for determining the absolute entropy of a substance or system. There are different ways of stating it. According to Max Planck, one way to state is that a perfect crystal has zero entropy at a temperature of absolute zero. A perfect crystal is one in which all of the atoms are aligned flawlessly with no defects of any kind in the crystalline structure. All of the atoms are motionless and have zero kinetic energy. This formulation is largely a statistical interpretation and involves the study of statistical mechanics, which is beyond the scope of this book. Nevertheless, it is a postulate, like the other two laws of thermodynamics, and its validity rests on experiment. However, since there are no means available to measure absolute entropies directly, it remains largely a theoretical principle.

Another common interpretation of the third law states that it is impossible to reach the state of absolute zero in a finite series of steps. This is entirely consistent with Planck's statement of the third law.

## 5.6   GIBBS FREE ENERGY AND THE SPONTANEITY OF A REACTION

The Gibbs free energy change, $\Delta G$, is defined as follows:

$$\Delta G = \Delta H - T \Delta S \qquad (5.11)$$

Its chief value is that it is an unequivocal measure and indication of a reaction's spontaneity. It includes both an enthalpy term, $\Delta H$, and an entropy term, $\Delta S$.

The driving force behind every chemical reaction is that it seeks simultaneously to reach a state of minimum energy and a state of maximum entropy. Thus, if the sign of $\Delta G$ is negative, the reaction proceeds spontaneously as written, though kinetically, it may be very slow; if positive, the *reverse* reaction proceeds spontaneously.

For equilibrium reactions, $\Delta G$ can be expressed as

$$\Delta G = -RT \ln K_c \text{ in general, for all reactions} \qquad (5.12)$$

$$\Delta G = -RT \ln K_p \text{ for gas} - \text{phase reactions} \qquad (5.13)$$

**Example 5.6:** (A & B) Gibbs Free Energy and Predicting
Reaction Spontaneity

Airborne, particulate sulfur, $S(s)$, may react with hydrogen gas, $H_2$ (g) to produce $H_2S$ gas, according to the following reaction:

$$H_2\left(g\right) + S(s) \rightarrow H_2S\left(g\right)$$

Given $\Delta H_f^\circ = -20.2\,\text{kJ/mol}$ and $\Delta S_f^\circ = +43.1\,\text{J/K} \cdot \text{mol}$.

A. Compute $\Delta G^\circ$ for the reaction.
B. Tell whether the reaction is spontaneous.

*Solution*

A. Use $\Delta G° = \Delta H° - T \Delta S°$. Note that $\Delta G°$ that will be computed is really $\Delta G_f°$. At standard conditions, $T = 298$ K, and all quantities are 1.0 mol.

$$\Delta G° = -20.2 \times 10^3 \text{ J} - (298 \text{ K})(+43.1 \text{ J/K}) = -20.2 \times 10^3 \text{ J} - 12.8 \times 10^3 \text{ J}$$

$$= -33,908 \text{ J} = -33.9 \text{ kJ}$$

B. Since the sign of $\Delta G$ is negative, the reaction is spontaneous as written.

**Important Note:** Note that the entropy term or value is smaller than the enthalpy term or value. This is normally the case for reactions at or below room temperature. The lower the temperature (measured in Kelvins), the smaller the contribution of the entropy term. Conversely, the higher the temperature of the reaction, the greater the contribution of the entropy term. In general, then, the enthalpy term plays a greater role in determining the spontaneity of a reaction.

**Example 5.7:** (A & B) Same Type As 5.6 (A & B )

At 1500°C, carbon monoxide reacts with hydrogen gas as shown in the reaction below. This is one way to synthesize methanol and eliminate carbon monoxide at the same time.

$$CO(g) + 2H_2(g) \rightleftharpoons CH_3OH(g)$$

It is known that $K_p = 1.4 \times 10^{-7}$.

A. Compute $\Delta G$ for the reaction.
B. State whether the reaction is spontaneous.

*Solution*

A. Use $\Delta G = -RT \ln K_p$. Then,

$$\Delta G = -(8.31 \text{ J/K} \cdot \text{mol})(1773 \text{ K}) \ln(1.4 \times 10^{-7})$$

$$= 233 \text{ J/K-mol}$$

B. Since the sign of $\Delta G$ is positive, this reaction is *not* spontaneous as written.

# 6 Chemical Kinetics

## 6.1 BASIC CONCEPT

**Chemical kinetics** is the study of reaction rates and mechanisms of chemical reactions. The reaction rate is a kinetic property and depends on the mechanism of the particular reaction. In contrast, thermodynamic properties are independent of mechanism. Thermodynamics indicates whether or not a reaction is spontaneous, but kinetics indicates how fast the reaction occurs and if it happens fast enough to be of any interest or value.

Consider the rusting of iron in the presence of air and water. Submerge an iron rod into a tank of water for 5 minutes and pull it out. It is highly unlikely that any rust will have formed and be evident. Thermodynamics states that iron will definitely react with air and oxygen and form rust (i.e., a family of hydrated iron oxide compounds). But kinetics says that it is a very slow reaction.

The easiest way to measure the rate of reactions is by the measuring the disappearance of a reactant, that is, the change in concentration of a reactant with time. Thus,

$$\text{Reaction rate} = \frac{-\Delta C}{\Delta t} = \frac{-(C_2 - C_1)}{t_2 - t_1}$$

where $\Delta C$ = the change in the concentration of the reactant and $\Delta t$ = the elapsed time. The subscripts 1 and 2 refer to concentrations and corresponding times, respectively. The reactant concentration decreases with time, making $(C_2 - C_1)$ a negative quantity. The minus sign is inserted in front of the parentheses to make it a positive quantity.

The reaction rate can be calculated just as well from the change in concentration with time of a product, in which case $\Delta C/\Delta t$ is a positive quantity. It is important to remember that the reaction rate is constantly changing, so $\Delta C/\Delta t$ represents an average rate. The instantaneous rate is given by the first derivative, $dC/dt$.

Thus, in very general terms, for the general reaction:

$$a\text{A} + b\text{B} \rightarrow g\text{G} + h\text{H} \tag{6.1}$$

the reaction rate may, in principle, be expressed as follows:

$$\text{Rate} = -\frac{1}{a}\frac{[A]}{\Delta t} = -\frac{1}{b}\frac{[B]}{\Delta t} = \frac{1}{g}\frac{[G]}{\Delta t} = \frac{1}{h}\frac{[H]}{\Delta t}$$

## 6.2 REACTION RATE LAWS, ORDERS, AND CONSTANTS

For the general reaction:

$$a\text{A} + b\text{B} \rightarrow g\text{G} + h\text{H} \tag{6.2}$$

DOI: 10.1201/9781003092759-6

the **rate law** or **rate equation** is more usefully defined and written as:

$$\text{Rate} = k[A]^x[B]^y \tag{6.3}$$

where $k$ = the **rate constant** (in reciprocal time and concentration units), [A] and [B] represent the molar concentrations of reactants A and B, [G] and [H] represent the molar concentrations of the products G and H, respectively, and $x$ and $y$ are exponents denoting the **order** of the reaction for the reactant species. The exponents $x$ and $y$ must be determined experimentally. The sum of the exponents $(x + y)$ is called the **overall order** of the reaction. In the present case, it is said that the reaction is $x$ order in A, $y$ order in B, and $(x + y)$ order overall. The order of reactions is generally zeroth, first, or second order, although in rare circumstances, either higher-order or fractional-order reactions are also possible. Most often, second-order reactions are ones where two identical molecules collide to react.

Consider, for example, the following reaction:

$$NO_2(g) + CO(g) \rightleftharpoons NO(g) + CO_2(g) \tag{6.4}$$

Experimentally, it is determined that the rate law for this reaction is

$$\text{Rate} = k[NO_2]^2 \tag{6.5}$$

This means that the rates are second order in $NO_2$, zeroth order in $CO_2$, and second order overall.

Consider, as another example, the decomposition of dinitrogen pentoxide, $N_2O_5$:

$$2N_2O_5(g) \rightleftharpoons 4NO_2(g) + O_2(g) \tag{6.6}$$

Here, it is experimentally determined that the rate law for this reaction is

$$\text{Rate} = k[N_2O_5] \tag{6.7}$$

Thus, the rate of decomposition is first order in $N_2O_5$. The rate constant $k$ is also experimentally measured:

$$k = 5.2 \times 10^{-3} \text{ at } 65°C$$

## 6.3  FIRST- AND SECOND-ORDER REACTIONS

For any reaction that is known to be first order in a particular species, as in the $N_2O_5$ decomposition discussed above, the rate can be written as

$$\text{Rate} = \frac{-d[A]}{dt} = \frac{-\Delta[A]}{\Delta t} = k[A] \tag{6.8}$$

**TABLE 6.1**

**Graphing Functions of Concentration Versus Time to Determine Reaction Order.**

| Order of Reaction | Linear [A] vs. Time Relationship |
|---|---|
| Zeroth order | [A] vs. $t$ |
| First order | $\ln[A]$ vs. $t$ |
| Second order | $1/[A]$ vs. $t$ |

In a **first-order reaction**, doubling the concentration of A means doubling the reaction rate. This equation may be integrated to give a more useful result:

$$\ln \frac{[A]_0}{[A]} = k\Delta t \tag{6.9}$$

This can also be rewritten in a more user-friendly way as follows:

$$\ln[A] = \ln[A]_0 - k\Delta t \tag{6.10}$$

In similar fashion, the integrated rate equation for a **second-order reaction** is found to be

$$\frac{1}{[A]} = \frac{1}{[A]_0} + k\Delta t \tag{6.11}$$

Note that $[A]_0$ represents the initial concentration of reactant A at time zero, while $[A]$ represents the concentration of A at any time $t$ during the reaction.

Table 6.1 summarizes ways to determine the order of a reaction by graphing concentration of reactant A versus time $t$.

## 6.4 HALF-LIFE OF A REACTION

A quantity defined as the half-life of a reaction can also be determined. The **half-life** is the time required for the concentration of the reactant to reach half of its initial value. This concept is encountered in many chemical reactions, nuclear chemistry and physics, and even in biology and pharmaceuticals. For a *first-order reaction*, Equation 6.10 can be reworked to show that the half-life is independent of the initial concentration and is given as follows:

$$t_{1/2} = \frac{0.693}{k} \tag{6.12}$$

For a *second-order reaction*, Equation 6.11 can be reworked to show that the half-life depends on initial concentration and is given as follows:

$$t_{1/2} = \frac{1}{k[A]_0} \tag{6.13}$$

Note that for second-order reactions, the half-life is concentration-dependent as can be seen in Equation 6.13. The variable $k$ is the rate constant, and $[A]_0$ is the initial concentration of the reactant A at time zero.

**Example 6.1:** (A & B) Determining the Rate Constant and Order of a Reaction

Consider the reaction between peroxydisulfate ion, $S_2O_3{}^{2-}$, and iodide ion, $I^-$, shown below and the experimental data about the concentrations and reaction rates at 25°C given in the table that follows.

$$S_2O_8{}^{2-}(aq) + 3I^-(aq) \rightarrow 2SO_4{}^{2-}(aq) + I_3{}^-(aq)$$

A. Write the rate law expression and determine the overall order, that is, $(x + y)$, for this reaction.
B. Determine the rate constant $k$ for this reaction at 25°C.

*Solution*

A. Examine the data given in Table 6.2. Comparing Experiment #1 with #2, it can be seen that doubling the iodide ion concentration, while holding the peroxydisulfate ion concentration constant, would double the reaction rate. Similarly, comparing Experiment #2 with #3 reveals that doubling the peroxydisulfate ion concentration, while holding the iodide ion concentration constant, would also double the reaction rate.
  Thus, this reaction is first order (or linear) in iodide ion concentration and first order (or linear) in peroxydisulfate ion concentration. The rate law can be written as follows:
  Rate $= k\,[S_2O_8{}^{2-}]^x[I^-]^y$
  where $x = 1$ and $y = 1$.
  Hence, Rate $= k[S_2O_8{}^{2-}]^1\left[I^-\right]^1$.
  The overall order is $(x + y) = (1 + 1) = 2$, indicating second order overall.
  This suggests that the mechanism for this reaction is **bimolecular**, indicating that the speed or rate of reaction depends on the concentrations of both the iodide and peroxydisulfate ions.
B. The rate constant $k$ can be determined using the data from *any* of the three experiments. Take the data from Experiment #1, and use the rate law expression determined in part (A). Thus,

---

**TABLE 6.2**

**Experimental Data for the Reaction Kinetics of Peroxydisulfate**

| EXPT. Number | $\left[S_2O_8{}^{2-}(aq)\right]$ | [I⁻(aq)] | Initial Rate |
|---|---|---|---|
| 1 | 0.080 M | 0.034 M | $2.2 \times 10^{-4}$ M/s |
| 2 | 0.080 M | 0.017 M | $1.1 \times 10^{-4}$ M/s |
| 3 | 0.16 M | 0.017 M | $2.2 \times 10^{-4}$ M/s |

---

$$2.2 \times 10^{(-4)} \text{ M/s} = k (0.034 \text{ M})^1 (0.080 \text{ M})^1$$

$$k = 8.09 \times 10^{-2} (\text{M} \times \text{s})^{-1}$$

## Example 6.2: (A & B) Understanding Half-Life and Computing Rate Constant

The thermal decomposition of phosphine $PH_3$ (produced by anaerobic bacteria and microbes and recently detected in the atmosphere of Venus) is known to be a first-order reaction:

$$4PH_3 (g) \rightarrow P_4 (g) + 6H_2 (g)$$

The half-life, $t_{1/2}$ is 35.0 s at 680°C.

A. Compute the rate constant $k$ for this reaction.
B. Find the time required for 75% of the initial concentration of $PH_3$ to decompose.

## Solution

A. For a first-order reaction, $t_{1/2} = 0.693/k$. Hence,

$$k = \frac{0.693}{t_{1/2}} = \frac{0.693}{35.0 \text{ s}} = 0.0198 \text{ s}^{-1}$$

B. To find the time, use the integrated, first-order rate law, Equation 6.9 or 6.10, and solve for $\Delta t$. If 75% of $PH_3$ is to decompose, 25% must remain! Thus,

$$[A]_0 = 100\% = 1.00$$

$$[A] = 25\% = 0.250$$

Either Equation 6.9 or 6.10 may be used. First, rearrange Equation 6.10 and solve for $\Delta t$:

$$k \Delta t = \ln[A]_0 - \ln[A]$$

$$k \Delta t = \{\ln[A]_0 - \ln[A]\}/k$$

Now substitute these values into Equation 6.10 and solve for $\Delta t$:

$$\Delta t = \{\ln(1.00) - \ln(0.250)\}/0.0198 \text{ s}^{-1}$$

$$\Delta t = 70.0 \text{ s}$$

## 6.5 DEPENDENCE ON TEMPERATURE: THE ARRHENIUS EQUATION

The Arrhenius model for rates of reactions assumes that molecules must undergo collisions before they can react. The number of collisions per unit time is **A**, the **frequency factor**. This model also assumes that not all collisions will result in or lead to a reaction. Rather, only those with sufficient energy, called the **activation energy, $E_a$** (to allow molecules to reach an **activated complex**) will achieve this result.

The rate constant $k$ is a function of temperature and can be expressed theoretically as the **Arrhenius Equation**:

$$k = Ae^{-E_a/RT} \tag{6.14}$$

where

$A$ = the frequency factor, a measure of the number of collisions per second,
$E_a$ = the activation energy in joules or calories per mole,
$R$ = the universal gas constant in units consistent with those for $E_a$,
$T$ = the absolute temperature in Kelvins.
A more useful form of the equation can be written as follows:

$$\ln k = \ln A - \frac{E_a}{RT} \tag{6.15}$$

The rate constants $k_1$ and $k_2$ at two different temperatures, $T_1$ and $T_2$, respectively, can be written as two separate equations, using Equation 2.24. The equations can be divided by each other and the result rearranged such that the rate constants are related as follows:

$$\ln\frac{k_1}{k_2} = \frac{E_a}{R}\left(\frac{1}{T_2} - \frac{1}{T_1}\right) \tag{6.16}$$

**Example 6.3:** (A & B) Using the Arrhenius Rate Equation

The reaction below between nitric oxide and ozone is important in the chemistry of air pollution as discussed in Chapter 12:

$$NO(g) + O_3(g) \rightarrow NO_2(g) + O_2(g)$$

The frequency factor $A = 8.70 \times 10^{12}\,s^{-1}$ and the rate constant $k = 300\,s^{-1}$ at 75°C.

    A. Find the activation energy, $E_a$, in joules per mole for this reaction.
    B. Find the rate constant $k$ of this reaction at 0°C, assuming $E_a$ to be constant.

*Solution*

    A. Use the Arrhenius Equation and solve for $E_a$.

$$\ln k = \ln A - \frac{E_a}{RT}$$

Use $R = 8.31$ J/K $\times$ mole and $T = 273 + 75°C = 348$ K. Then,

$$\ln 300 = \ln 8.70 \times 10^{12} - \frac{E_a}{(8.31)(348)}$$

Thus, $E_a = 69{,}700$ J/mol or 69.7 kJ/mol

B. Denote the rate constant given at 75°C as $k_1$, and the rate constant at 0°C as $k_2$, and use Equation 6.16:

$$\ln \frac{k_1}{k_2} = \frac{E_a}{R}\left(\frac{1}{T_2} - \frac{1}{T_1}\right)$$

Substitute the above information, and solve for $k_2$.

$$\ln k_2 = \ln 300 - \frac{69{,}700}{8.31}\left(\frac{1}{273} - \frac{1}{348}\right)$$

$$= 5.7038 - 6.6210 = -0.9172$$

$$k_2 = 0.3996\,\text{s}^{-1} = 0.400\,\text{s}^{-1}$$

## 6.6   CATALYSIS

Substances that increase the rate of reaction without themselves being consumed in the reaction are called **catalysts**. They work by lowering the activation energy, $E_a$, the barrier required for the reaction to proceed. A catalyst may be in the same physical state as the other species in the reaction (homogeneous catalysis) or in a different physical state (heterogeneous catalysis).

# 7 Electrochemistry and RED-OX Reactions

## 7.1 BASIC CONCEPT

One category of chemical reactions described briefly in Section 1.9.2 is oxidation–reduction reactions or "red-ox" reactions for short. Oxidation–reduction ("red-ox") reactions are reactions in which one substance is oxidized while another is simultaneously reduced. The processes of oxidation and reduction can be defined as follows. Oxidation is the loss of electrons, while reduction is the gain of electrons. However, diagnostically speaking, a substance is oxidized or reduced, respectively, if any *one* of the following conditions is met:

Oxidation

- Loss of electrons
- Gain of oxygen atoms
- Loss of hydrogen atoms

Reduction

- Gain of electrons
- Loss of oxygen atoms
- Gain of hydrogen atoms

Consider the arrangement shown in Figure 7.1. This arrangement is known as an **electrochemical cell**, and the underlying electrochemical reaction involves simultaneous oxidation and reduction reactions. Electrochemical cells are the basis of the operation of all batteries. If the reactions and current flow are spontaneous, as in any battery, the cell is called a **voltaic** or **galvanic cell**. If, instead, electrical energy must be supplied, the cell is called an **electrolytic cell**, and the process is known as **electrolysis**.

An electrochemical cell is comprised of two **half-cells**. In one, called the **anode**, **oxidation** occurs; in the other, called the **cathode**, **reduction** occurs. Each reaction is called a **half-reaction**. The sum of the two half-reactions is the overall or net reaction. The rusting of iron or the corrosion of any metal involves an oxidation–reduction process, with a net electromotive potential.

Some useful physical measurements and their units in electrochemical cells include the volt (V), the electron volt (eV), the coulomb (C), and the ampere (A).

Since 1 volt = 1 joule per coulomb, and 1 ampere = 1 coulomb per second, then (current in amperes) $\times$ (time in seconds) = total charge transferred in coulombs.

DOI: 10.1201/9781003092759-7

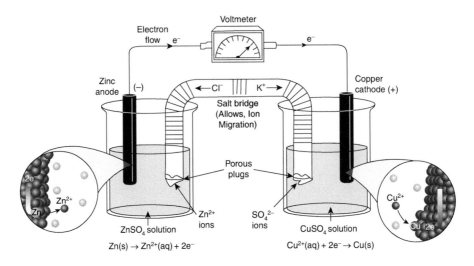

**FIGURE 7.1**    A copper–zinc electrochemical galvanic cell.

In addition, (potential in volts) × (charge in coulombs) = total energy produced or used in joules. See Equations 7.7–7.13 for a summary of useful quantitative relationships for electrochemical cells.

## 7.2    THE NERNST EQUATION

The electromotive potential developed by the cell diagramed in Figure 2.3 is very useful. It can be harnessed and can serve as the basis of battery power. In fact, pairs of different metals can be similarly arranged to produce a variety of desired voltages.

When all substances used in the cell are in their standard states, the net voltage is the simple sum of the individual cell potentials or voltages listed in Table 2.4. Standard states means that all metals are pure, their starting salt concentrations in solution are 1.00 molar, and the ambient or operating temperature is 25°C. Such a situation is outlined in the section labeled **A Simple Electrochemical Cell** below, using the diagram in Figure 2.3. A specific set of conditions is given in Example 2.28.

However, when nonstandard conditions exist, another equation must be used to calculate the net potential of the electrochemical cell. Nonstandard conditions usually refer to operating temperatures other than 25°C or concentration values of dissolved metal species other than 1.00 molar. The law describing the relationship between the concentrations of ionic solutions and the voltage produced in the net reaction is called the **Nernst Equation** and is expressed as follows:

$$E_{net} = E_{net}{}^{\circ} - \frac{RT}{nF}\ln Q \qquad (7.1)$$

where

$E^{\circ}_{net}$ = the total or net cell potential in volts obtained by the addition of the two
potentials corresponding to the two standard half-cell reactions under
standard conditions.

$E_{net}$ = the total or net cell potential in volts of the cell operating under nonstandard conditions, usually due to a temperature other than 25°C, or concentration values of dissolved metal species other than 1.00 molar.

$R$ = the gas constant = 8.31 J/K · mol.

$T$ = the absolute temperature in kelvins.

$F$ = the Faraday constant = 96,485 C/mol electrons.

$n$ = the number of moles of electrons transferred per unit reaction.

$Q$ = the ratio of molar concentrations of products to molar concentrations of reactants, for all (aq) or (g) species in solution [species that have an (s) or (l) state are excluded]. See the earlier discussion on equilibrium concepts in Chapter 4.1.

Note that the term "$E_{net}$" may be referred to as either the cell potential or the electromotive force potential or simply as the EMF of the cell.

For the condition where $T = 25°C = 298K$, and $\ln Q = 2.303 \log Q$, the Nernst Equation can be simplified to

$$E_{net} = E_{net}° - \frac{0.0591}{n} \log Q \qquad (7.2)$$

This fact can produce seemingly endless combinations of cell arrangements and variations, and generate exactly desired voltages for very specific applications. Note that when standard conditions are met, the log $Q$ term vanishes, and the Nernst Equation simplifies to $E_{net} = E°_{net}$. An example of this type of problem is given in Example 7.1.

Table 7.1 shows the standard reduction reactions of metals, along with selected nonmetals, and their corresponding potentials, relative to the standard hydrogen electrode (whose reduction potential is taken as zero) at 25°C. More extensive tables of reduction potentials organized according to acidic and alkaline conditions can be found in the Handbook of Chemistry and Physics. It is important to note that reduction reactions and potentials depend on the pH of solution, as well as the presence of any complexing agents. This listing is often referred to as the **electromotive series**, and the reduction potentials are given as the $E°$ values in volts. If the $E°$ value listed is positive, the reaction is spontaneous as written. If negative, the reverse reaction is spontaneous. The superscript zero refers to standard conditions, that is, a temperature of 25°C, a concentration of 1.00 M for any ions present in solution, and a partial pressure of 1.00 atmosphere for any gases present.

The listing is also referred to as the **activity series** of metals, since they indicate which metal is more likely, relative to another, to release its electrons, oxidize, and react. The reactions posted are sometimes referred to as **half-cell reactions**. For example, the metal lithium has the highest *negative* $E°$ value, −3.045 V, while magnesium has an $E°$ value of −2.363 V. Therefore, elemental lithium will oxidize and react more easily than elemental magnesium. The nonmetal fluorine has the highest *positive* $E°$ value, +2.87 V, and is thus easiest to reduce, In other words, a large negative $E°$ value means that the reactant is a good reducing agent, and a large positive $E°$ value indicates a good oxidizing agent.

**TABLE 7.1**

**The Electromotive Series Expressed as Standard Reduction Potentials at STP**

| E° (Volts) | Half-Cell Reaction |
|---|---|
| +2.87 | $F_2(g) + 2e^- \rightleftharpoons 2F^-(aq)$ |
| +2.08 | $O_2(G) + 2H^+(aq) + 2e^- \rightleftharpoons O_2(g) + H_2O$ |
| +2.05 | $S_2O_8{}^{2-}(aq) + 2e^- \rightleftharpoons 2SO_4{}^{2-}(aq)$ |
| +1.82 | $Co^{3+}(aq) + e^- \rightleftharpoons Co^{2+}(aq)$ |
| +1.77 | $H_2O_2(aq) + 2H^+(aq)\ 2e^- \rightleftharpoons 2H_2O$ |
| +1.695 | $MnO_4{}^-(aq) + 4H^+(aq) + 3e^- \rightleftharpoons MnO_2(s) + 2H_2O$ |
| +1.69 | $PbO_2(s) + SO_4{}^{2-}(aq) + 4H^+(aq) + 2e^- \rightleftharpoons PbO_4(s) + 2H_2O$ |
| +1.63 | $2HCOCl(aq) + 2H^+(aq) + e^- \rightleftharpoons Cl_2(g) + 2H_2O$ |
| +1.51 | $Mn^{3+}(aq) + e^- \rightleftharpoons Mn^{2+}(aq)$ |
| +1.49 | $MnO_4{}^-(aq) + 8H^+(aq) + 5e^- \rightleftharpoons Mn^{2+}(aq) + 4H_2O$ |
| +1.46 | $PbO_2(s) + 4H^+(aq) + 2e^- \rightleftharpoons Pb^{2+}(aq) + 2H_2O$ |
| +1.44 | $BrO_3{}^-(aq) + 6H^+(aq) + 6e^- \rightleftharpoons Br^-(aq) + 3H_2O$ |
| +1.42 | $Au^{3+}(aq) + 3e^- \rightleftharpoons Au(s)$ |
| +1.36 | $Cl_2(g) + 2e^- \rightleftharpoons 2Cl^-(aq)$ |
| +1.33 | $Cr_2O_7{}^{2-}(aq) + 14H^+(aq) + 6e^- \rightleftharpoons 2Cr^{3+}(aq) + 7H_2O$ |
| +1.24 | $O_3(g) + H_2O + 2e^- \rightleftharpoons O_2(g) + 2OH^-(aq)$ |
| +1.23 | $MnO_2(s) + 4H^+(aq) + \rightleftharpoons Mn^{2+}(aq) + 2H_2O$ |
| +1.23 | $O_2(g) + 4H^+ (aq) + 4e^- \rightleftharpoons 2H_2O$ |
| +1.20 | $Pt^2(aq) + 2e^- \rightleftharpoons Pt(s)$ |
| +1.07 | $Br_2(aq) + 2e^- \rightleftharpoons 2Br^-(aq)$ |
| +0.96 | $NO_3{}^-(aq) + 4H^+(aq) + 3e^- \rightleftharpoons NO(g) + 2H_2O$ |
| +0.94 | $NO_3{}^-(aq) + 3H^+(aq) + 2e^- \rightleftharpoons HNO_2(aq) + H_2O$ |
| +0.91 | $2Hg^{2+}(aq) + 2e^- \rightleftharpoons Hg_2{}^{2+}(aq)$ |
| +0.87 | $HO_2{}^-(aq) + H_2O + 2e^- \rightleftharpoons 2NO_2(g) + 2H_2O$ |
| +0.80 | $NO_3{}^-(aq) + 4H^+(aq) + 2e^- \rightleftharpoons 2NO_2(g) + 2H_2O$ |
| +0.80 | $Ag^+(aq) + e^- \rightleftharpoons Ag(s)$ |
| +0.77 | $Fe^{3+}(aq) + e^- \rightleftharpoons Fe^{2+}(aq)$ |
| +0.69 | $O_2(g) + 2H^+(aq) + 4e^- \rightleftharpoons H_2O_2(aq)$ |
| +0.54 | $I_2(s) + 2e^- \rightleftharpoons 2I^-(aq)$ |
| +0.49 | $NiO_2(s) + 2H_2O + 2e^- \rightleftharpoons Ni(OH)_2(s) + 2OH^-(aq)$ |
| +0.45 | $SO_2(aq) + 4H^+ (aq)\ 4e^- \rightleftharpoons S(s) + 2H_2O$ |
| +0.401 | $O_2(g) + 2H_2O + 4e^- \rightleftharpoons 4OH^-(aq)$ |
| +0.34 | $Cu^{2+}(aq) + 2e^- \rightleftharpoons Cu(s)$ |
| +0.27 | $Hg_2Cl_2(s) + 2e^- \rightleftharpoons 2Hg(1) + 2Cl^-(aq)$ |
| +0.25 | $PbO_2(s) + H_2O + 2e^- \rightleftharpoons PbO(s) + 2OH^-(aq)$ |
| +0.2223 | $AgCl(s) + e^- \rightleftharpoons Ag(s) + Cl^-(aq)$ |
| +0.172 | $SO_4{}^{2-}(aq) + 4H^+(aq) + 2e^- \rightleftharpoons H_2SO_3(aq) + H_2O$ |
| +0.169 | $S_4O_6{}^{2-}(aq) + 2e^- \rightleftharpoons 2S_2O_3{}^{2-}(aq)$ |
| +0.16 | $Cu^{2+}(aq) + e^- \rightleftharpoons Sn^{2+}(aq)$ |

*(Continued)*

**TABLE 7.1** (*Continued*)
**The Electromotive Series Expressed as Standard Reduction Potentials at STP**

| E° (Volts) | Half-Cell Reaction |
|---|---|
| +0.15 | $Sn^{4+}(aq) + 2e^- \rightleftharpoons H_2S(g)$ |
| +0.14 | $S(s) + 2H^+(aq) + 2e^- \rightleftharpoons H_2S(g)$ |
| +0.07 | $AgBr(s) + e^- \rightleftharpoons Ag(s) + Br^-(aq)$ |

| E (Volts) | Half-Cell Reaction |
|---|---|
| 0.00 | $2H^+(aq) + 2e^- \rightleftharpoons H_2(g)$ |
| −0.13 | $Pb^2(aq) + 2e^- \rightleftharpoons H_2(g)$ |
| −0.14 | $Sn^{2+}(aq) + 2e^- \rightleftharpoons Sn(s)$ |
| −0.15 | $AgI(s) + e^- \rightleftharpoons Ag(s) + I^-(aq)$ |
| −0.25 | $Ni^{2+}(aq) + 2e^- \rightleftharpoons Ni(s)$ |
| −0.28 | $Co^{2+}(aq) + 2e^- \rightleftharpoons Co(s)$ |
| −0.34 | $In^{3+}(aq) + 3e^- \rightleftharpoons In(s)$ |
| −0.34 | $Tl^+(aq) + e^- \rightleftharpoons Tl(s)$ |
| −0.36 | $PbSO_4(s) + 2e^- \rightleftharpoons Pb(s) + SO_4^{2-}(aq)$ |
| −0.40 | $Cd^{2+}(aq) + 2e^- \rightleftharpoons Cd(s)$ |
| −0.44 | $Fe^{2+}(aq) + 2e^- \rightleftharpoons Fe(s)$ |
| −0.56 | $Cr^{3+}(aq) + 3e^- \rightleftharpoons Ga(s)$ |
| −0.58 | $Zn^{2+}(aq) + 2e^- \rightleftharpoons Zn(s)$ |
| −0.74 | $Cd(OH)_2(s) + 2e^- \rightleftharpoons Cd(s) + 2OH^-(aq)$ |
| −0.76 | $2H_2O + 2e^- \, 2H_2O + 2e \, Zn(s)$ |
| −0.81 | $Cd(OH)_2(s) + 2e^- \, 2H_2O + 2e \rightleftharpoons Cd(s) + 2OH^-(aq)$ |
| −0.83 | $2H_2O + 2e^- \rightleftharpoons H_2(g) + 2OH^-(aq)$ |
| −0.88 | $Fe(OH)_2(s) + 2e^- \rightleftharpoons Cd(s) + 2OH^-(aq)$ |
| −0.91 | $Cr^{2+}(aq) + e^- \rightleftharpoons Cr(s)$ |
| −1.16 | $N_2(g) + 4H_2O + 4e^- \rightleftharpoons N_2O_4(aq) + 4OH^-(aq)$ |
| −1.18 | $V^{2+}(aq) + 2e^- \rightleftharpoons V(s)$ |
| −1.216 | $ZnO_2^-(aq) + 2H_2O + 2e^- \rightleftharpoons Zn(s) + 4OH^-(aq)$ |
| −1.63 | $Ti^{2+}(aq) + 2e^- \rightleftharpoons Ti(s)$ |
| −1.66 | $Al^{3+}(aq) + 2H_2O + 2e^- \rightleftharpoons Al(s)$ |
| −1.79 | $U^{3+}(aq) + 3e^- \rightleftharpoons U(s)$ |
| −2.02 | $Sc^{3+}(aq) + 3e^- \rightleftharpoons Sc(s)$ |
| −2.36 | $La^{3+}(aq) + 3e^- \rightleftharpoons La(s)$ |
| −2.37 | $Y^{3+}(aq) + 3e^- \rightleftharpoons Y(s)$ |
| −2.37 | $Mg^{2+}(aq) + 2e^- \rightleftharpoons Mg(s)$ |
| −2.89 | $Na^+(aq) + e^- \rightleftharpoons Na(s)$ |
| −2.90 | $Ca^{2+}(aq) + e^- \rightleftharpoons Ca(s)$ |
| −2.92 | $Sr^{3+}(aq) + e^- \rightleftharpoons Sr(s)$ |
| −2.93 | $Ba^{2+}(aq) + e^- \rightleftharpoons Cs(s)$ |
| −3.05 | $Li^+(aq) + e^- \rightleftharpoons Li(s)$ |

## 7.3 A SIMPLE ELECTROCHEMICAL CELL

Consider the copper/zinc electrochemical cell shown in Figure 7.1.

In the right half-cell, the cathode, $Cu^{2+}$ ions in solution are being reduced to $Cu^0$, indicating solid copper metal, deposit on the Cu cathode. In the left half-cell, the anode, $Zn^0$ or solid zinc is being oxidized to $Zn^{2+}$ ions, which dissolve in solution. The anions—$SO_4^{2-}$ (aq)—are spectator ions and do not participate in the reaction. The salt bridge is necessary to maintain electrical neutrality and retard polarization.

The shorthand notation for this cell is as follows:

$$Zn(s) \left| ZnSO_4(aq)(1.00 \text{ M}) \right\| CuSO_4(aq)(1.00 \text{ M}) \left| Cu(s) \right. \qquad (7.3)$$

It is understood that the anode cell, where oxidation occurs, is written first and is separated by a double vertical line from the cathode cell, where reduction occurs. Single vertical lines separate the solid electrode from the aqueous solution into which it is immersed. Concentrations of the solutions are expressed in moles per liter in parentheses.

The two half-reactions, followed by the **net reaction**, can be written as follows:

At cathode

$$Cu^{2+}(aq) + 2e^- \rightarrow Cu^0(s) \qquad E^\circ = 0.337 \text{ V} \qquad (7.4)$$

At anode

$$Zn^0(s) \rightarrow Zn^{2+}(aq) + 2e^- \qquad E^\circ = 0.763 \text{ V} \qquad (7.5)$$

Net reaction

$$Cu^{2+}(aq) + 2e + Zn^0(s) \rightarrow Cu^0(s) + Zn^{2+}(aq) + 2e \qquad E_{net}^\circ = 1.100 \text{ V}$$

Or finally,

$$Cu^{2+}(aq) + Zn^0(s) \rightarrow Cu^0(s) + Zn^{2+}(aq) \qquad E_{net}^\circ = 1.100 \text{ V} \qquad (7.6)$$

Note that the electrons exactly cancel each other in the two half-reactions and do not appear in the *final* net reaction. If this cancellation of electrons does not happen automatically, one or both of the half-reactions must be multiplied by a suitable coefficient (i.e., an integer) or coefficients to obtain the exact cancellation of electrons. Note also that the positive value for $E_{net}^\circ$ indicates that this reaction proceeds **spontaneously** and the cell produces 1.100 volts under standard conditions.

Also note that the $E^\circ$ value at the cathode represents a reduction potential and may be written as $E^\circ_{red}$, while the $E^\circ$ value at the anode represents an oxidation potential and may be written as $E^\circ$ox. This convention may be used in other reference manuals or textbooks.

## 7.3.1   Useful Quantitative Relationships

$$1 \text{ mole of electrons} \left( e^- \right) = 96,485 \text{ coulombs I} \tag{7.7}$$

$$\text{charge I} = \text{current} \left( C/s \right) \times \text{time} \left( s \right) \tag{7.8}$$

$$1 \text{ ampere} \left( A \right) = 1 \text{ coulomb} / 1 \text{ second} \tag{7.9}$$

$$1 \text{ volt} \left( V \right) = 1 \text{ joule} \left( J \right) / 1 \text{ coulomb} \tag{7.10}$$

$$\text{Faraday constant } F = 96,485 \text{ coulombs/mole } e^- \tag{7.11}$$

$$\text{Gas constant } R = 8.31 \text{ joules/Kelvin} \cdot \text{mole} \tag{7.12}$$

$$\text{Number of moles of } e^- = \left( \text{current} \times \text{time} \right) / F \tag{7.13}$$

**Example 7.1:** EMF Potential of a Cell/Battery Under Standard Conditions

Consider the electrochemical cell with the following net reaction, which is observed to proceed spontaneously. All species are in their standard states, i.e., concentrations of dissolved species are 1.00 molar, and $T = 25°C$.

$$Mg(s) + Sn^{2+}(aq) \rightarrow Mg^{2+}(aq) + Sn(s)$$

A. Using Table 7.1, write the two half-reactions with their respective $E°$ values.
B. Indicate which species is oxidized and which is reduced.
C. Identify the anode and cathode.
D. Compute $E_{net}°$.
E. Confirm that the reaction proceeds spontaneously.

### Solution

A. $Mg(s) \rightarrow Mg^{2+}(aq) + 2e^- \quad\quad E° = 2.370 \text{ V}$

   $Sn^{2+}(aq) + 2e^- \rightarrow Sn(s) \quad\quad E° = -0.140 \text{ V}$

B. $Mg(s)$ is oxidized to $Mg^{2+}(aq)$, while $Sn^{2+}(aq)$ is reduced to $Sn(s)$.
C. The anode is the electrode where oxidation takes place, while the cathode is the electrode where reduction takes place. Since Mg is oxidized, the half-cell containing the Mg electrode and $Mg^{2+}(aq)$ solution must be the anode.

   Similarly, the cathode is the half-cell containing the Sn electrode dipped into $Sn^{2+}(aq)$ solution.
D. Adding the two half-cell potentials determined in part **A** gives $E_{net}° = 2.230 \text{ V}$.
E. Since $E_{net}°$ is a positive number, the net reaction must proceed spontaneously.

**Example 7.2:** EMF Potential of a Cell/Battery Under Non-Standard Conditions

Consider the reaction in Example 7.1. Instead of standard conditions of 1.00 M concentrations for each solution, assume now that the $Mg^{2+}$(aq) solution is 0.850 M and that the $Sn^{2+}$(aq) solution is 0.0150 M. Find the $E_{net}°$ under these conditions, assuming $T = 25.0°C$.

### Solution

Since this is under nonstandard conditions, the Nernst Equation applies and must be used. However, use the simplified version of this equation, Equation 7.2, since $T = 25.0°C$.

$$E_{net} = E_{net}° - \frac{0.0591}{n} \log Q$$

In this case, $n = 2$, since 2 mol of electrons is exchanged in the *net* reaction. Also, by examining the net reaction, $Q$ is found to be

$$Q = \frac{\left[ Mg^{2+}(aq) \right]}{\left[ Sn^{2+}(aq) \right]} = \frac{0.850\,M}{0.0150\,M}$$

$E_{net}°$ was calculated in part $D$ of Example as 2.230 V. Substitution now gives

$$E_{net} = 2.230\,V - \frac{0.0591}{2} \log \frac{0.850\ M}{0.0150\ M}$$

$$= 2.178\,V$$

**Important Note:** It is important to note that this answer is lower than the answer in Example 7.1, which uses the same metal electrodes. Thus, this demonstrates that the galvanic cell voltage (EMF) may be adjusted, i.e., either increased or decreased, by changing the concentration of either the oxidizing or reducing agent or both. If one of these is a hydrogen ion or hydroxide ion, this means changing the pH of one of the cell solutions.

**Example 7.3:** Calculating the Gibbs Free Energy Change for an EMF System

Compute the Gibbs free energy change, $\Delta G$, for the cell in Example 7.2.

### Solution

$$\Delta G = nFE_{net}$$

$$= -\left( 2\,mol\,e^- \right)\left( 96,485\,C/mol\,e^- \right)\left( 2.178\,V \right)$$

$$= -420,290\,J \text{ or } -420.3\,kJ$$

Note that the minus sign indicates energy released!

**Example 7.4:** (A-C) Evaluating an EMF System for Total Potential and Time

Consider the following cell reaction, in which all species are standard-state conditions:

$$Cu^{2+}(aq) + H_2(g) \rightarrow Cu(s) + 2H^+(aq)$$

A. Predict the effect on the electromotive potential of this cell of adding NaOH solution to the hydrogen half-cell until pH = 7.
B. Compute the number of coulombs required to deposit 4.20 g of Cu(s) in the copper half-cell.
C. How long, in seconds, will this deposition take, if the measured current is 4.00 A?

## Solution

A. Refer to the simplified Nernst Equation, Equation 7.2. The EMF is the $E_{net}$:

$$E_{net} = E_{net}° - \frac{0.0591}{2} \log \frac{\left[H^+\right]^2}{\left[Cu^{2+}\right]}$$

If NaOH(aq) is added, $H^+$ ions (aq) will be neutralized, thereby raising the pH. $[H^+(aq)]$ will decrease, reducing the magnitude of the log term, which is subtracted from $E_{net}°$. This, in turn, will increase the $E_{net}$ or the EMF of the cell.

B. First, compute the number of electrons that must be transferred to deposit 4.20 g of Cu(s). Then, recall that 1.00 mol of $Cu^{2+}$ (aq) is deposited as Cu(s) for every 2.0 mol of electrons used and that the Faraday constant $F = 96,485$ C/1.00 mol electrons. Thus,

$$4.20 \, g \, Cu(s)\left(\frac{1.00 \, mol \, Cu}{63.5 \, g \, Cu}\right) = 0.06614 \, mol \, Cu(s) \, deposited$$

$$0.06614 \, mol \, Cu(s)\left(\frac{2 \, mol \, e^-}{1.00 \, mol \, Cu(s)}\right) = 0.1323 \, mol \, e^-$$

$$0.1323 \, mol \, e^-\left(\frac{96,485 \, C}{1.00 \, mol \, e^-}\right) = 1.28 \times 10^4 \, C$$

$$Time, t \, (s) = \frac{charge \, (C)}{current \, \left(C/s\right)}$$

$$= \frac{1.28 \times 10^4 \, C}{4.00 \, A}$$

$$= 3.20 \times 10^3 \, s$$

This is about 53 min or a little less than 1.0 h.

# 8 Organic Chemistry
## *Naming, Structure, and Isomerism*

**Organic Chemistry** is the broad branch of chemistry that focuses on the study of compounds that are composed mainly of carbon and hydrogen. There are tens of millions of organic compounds known, both natural and synthetic, due primarily to the fact that carbon has the rather unique ability to bond to itself. With the exception of carbon dioxide, the carbonate and bicarbonate ions, if a compound contains carbon, it is considered to be an organic compound. If the Chemical Abstract Services Registry (CASR), which catalogs all known chemical substances in the marketplace today, is examined, it is readily apparent that the overwhelming majority are organic compounds—an approximate total of 167 million as of this publication date. It is useful to note that every known chemical substance has a unique CAS registration number, i.e., CASRN, for identification purposes and a list of all of its known properties.

## 8.1 ORGANIC COMPOUNDS COMPARED TO INORGANIC COMPOUNDS

The vast number of organic compounds can be attributed to the special nature of carbon and its location in the periodic table. These contributors are as follows:

- The ability of carbon to form up to four covalent bonds.
- The ability of carbon atoms to link to each other to form seemingly endless, elongated chains. Links can be linear or branched out—leading to molecules with the same molecular formula but different connectivity, i.e., constitutional isomers (illustrated in Figure 8.5 and explained in Section 8.6 in this chapter). This ability also leads to the formation of polymers, which are discussed in more detail in Chapter 10.
- The ability of carbon to link in circles or polygons. These cyclic compounds can additionally fuse to form complex molecules.
- C–C bonds can be single, double, or triple.
- Substituents, i.e., individual atoms or groups of atoms replacing hydrogen atoms, may have different orientations in space leading to stereoisomeric compounds (explained in Section 8.6 in this chapter and Section 9.1 in the next chapter).
- The ability of carbon to bond to other atoms besides hydrogen, e.g., oxygen, nitrogen, sulfur, phosphorous, and the halides (fluorine, chlorine, bromine, and iodine).

DOI: 10.1201/9781003092759-8

## 8.2   THE NEED FOR KNOWLEDGE OF ORGANIC CHEMISTRY

The fact that there are millions of organic molecules means that organic products are encountered in everyday life more often than realized. The study of organic chemistry impacts life in many ways. Vital life molecules such as amino acids, enzymes, proteins, steroids, nucleic acids, and carbohydrates all contain carbon. Understanding how these molecules work requires organic chemistry knowledge.

Organic chemistry is the backbone of the pharmaceutical industry. The story of any medicine—from discovery and isolation, to synthesis and characterization, to its safety and efficacy—involves organic chemistry.

Organic molecules also make our foods, whether plant or animal based. While the term "organic foods" is commonly used to imply agricultural products that are grown without the use of synthetic chemicals (pesticides) and bioengineered genes, it is important to note that any living thing (or its product thereof) contains C–H bonds (or hydrocarbons) and is, hence, organic.

Our clothing (the textile industry), fuels, energy, plastics, soaps/detergents, and solvents are also organic. Acetone, for example, is a common organic solvent in industrial settings or processes, laboratories, and at home. Gasoline is mainly a mixture of C–H or hydrocarbon containing compounds.

Organic molecules (e.g., graphene) are also gaining widespread use in the electronics industry to make nanoscale switches while others are being explored for use as cancer treatment drug carriers. Additionally, organic chemistry knowledge is vital for the resolution of environmental issues, such as identification of hazardous materials and climate change. A lot of research, with some very promising results, is currently underway for alternative fuel sources, for example.

Some basic organic chemistry knowledge is also necessary for one's health and safety. Just like most chemicals, a slight difference in elemental composition of organic compounds can mean a dramatic difference in reactivity. For example, methanol ($CH_3OH$) and ethanol ($CH_3CH_2OH$) are two organic compounds that belong to the alcohol family. Ethanol is a key component of alcoholic beverages and is relatively safe, if consumed in small quantities. Methanol, on the other hand, if consumed even in small amounts can lead to blindness or even death.

## 8.3   HYDROCARBONS: ALIPHATIC AND
##         AROMATIC HYDROCARBONS

Organic compounds composed of carbon and hydrogen only are known as hydrocarbons. Hydrocarbons with no benzene group/ring are known as aliphatic hydrocarbons while those with benzene group/ring are known as aromatic hydrocarbons (Figure 8.1).

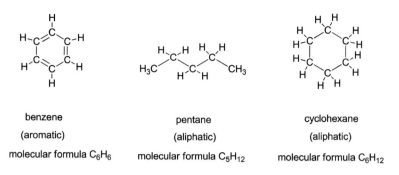

benzene
(aromatic)
molecular formula $C_6H_6$

pentane
(aliphatic)
molecular formula $C_5H_{12}$

cyclohexane
(aliphatic)
molecular formula $C_6H_{12}$

**FIGURE 8.1** Common hydrocarbons.

**Aliphatic Hydrocarbons** are further divided into alkanes, alkenes, and alkynes, depending on C–C bond types, i.e., single, double, or triple bonds, respectively.

### 8.3.1 ALKANES

Alkanes are the simplest hydrocarbons with the general formula $C_nH_{2n+2}$ where n is a positive, nonzero integer. They are made entirely of C–C single covalent bonds and are therefore said to be saturated because they contain the maximum number of hydrogen atoms that can bond with the number of carbon atoms present.

Alkanes are used as solvents, fuels, and starting materials for synthesis of other organic molecules. Methane ($CH_4$), which is obtained commercially from natural gas, is the simplest alkane. Propane, the third alkane, is used for cooking, water and space heating, to power buses, forklifts, etc.

Table 8.1 lists the first ten alkanes, comprising what is often called a "homologous series."

**TABLE 8.1**
**Common Alkanes**

| Alkane Name | Molecular Formula | Alkane Name | Molecular Formula |
|---|---|---|---|
| Methane | $CH_4$ | Hexane | $C_6H_{14}$ |
| Ethane | $C_2H_6$ | Heptane | $C_7H_{16}$ |
| Propane | $C_3H_8$ | Octane | $C_8H_{18}$ |
| Butane | $C_4H_{10}$ | Nonane | $C_9H_{20}$ |
| Pentane | $C_5H_{12}$ | Decane | $C_{10}H_{22}$ |

### 8.3.2 ALKENES

Alkenes, also known as olefins, are hydrocarbons containing at least one carbon–carbon double bond. They have the general formula $C_nH_{2n}$ where n is a positive integer, greater than 1. Alkenes are said to be unsaturated hydrocarbons as up to two hydrogen atoms can be added across a double bond in a chemical reaction.

Alkenes are used as synthetic precursors for alcohols, polymers, etc. Ethene ($C_2H_4$), also known as ethylene, is the simplest alkene. It is a significant factor in plant growth and other transformations such as ripening of fruits, seed germination, and maturation of flowers. Ethene molecules can undergo radical addition reactions to generate polyethylene polymer, also known as polyethene. Table 8.2 lists the first ten alkenes.

**TABLE 8.2**
**Common Alkenes**

| Alkene Name | Molecular Formula | Alkene Name | Molecular Formula |
|---|---|---|---|
| Ethene | $C_2H_4$ | Heptene | $C_7H_{14}$ |
| Propene | $C_3H_6$ | Octene | $C_8H_{16}$ |
| Butene | $C_4H_8$ | Nonene | $C_9H_{18}$ |
| Pentene | $C_5H_{10}$ | Decene | $C_{10}H_{20}$ |
| Hexene | $C_6H_{12}$ | Undecene | $C_{11}H_{22}$ |

### 8.3.3 ALKYNES

Alkynes are hydrocarbons containing at least one carbon–carbon triple bond. They have the general formula $C_nH_{2n-2}$ where n is a positive integer, greater than 1. Alkynes are said to be unsaturated hydrocarbons as up to four hydrogen atoms can be added across a triple bond in a chemical reaction.

Alkynes are used as synthetic precursors for simple organic molecules as well as polymers, among other things. Ethyne ($C_2H_2$), also known as acetylene, is the simplest alkyne. Ethyne is used industrially for welding. Table 8.3 lists the first ten alkynes.

**TABLE 8.3**
**Common Alkynes**

| Alkyne Name | Molecular Formula | Alkyne Name | Molecular Formula |
|---|---|---|---|
| Ethyne | $C_2H_2$ | Heptyne | $C_7H_{12}$ |
| Propyne | $C_3H_4$ | Octyne | $C_8H_{14}$ |
| Butyne | $C_4H_6$ | Nonyne | $C_9H_{16}$ |
| Pentyne | $C_5H_8$ | Decyne | $C_{10}H_{18}$ |
| Hexyne | $C_6H_{10}$ | Undecyne | $C_{11}H_{20}$ |

### 8.3.4 AROMATIC HYDROCARBONS

Aromatic hydrocarbons contain one or more benzene (Figure 8.2) rings. Benzene rings, despite having multiple double bonds, are known for their unusual stability and only undergo addition and substitution reactions under special conditions. Three molecules of hydrogen, for example, can be readily added across three isolated double bonds in a molecule. Addition of three molecules of hydrogen to benzene ring, on the other hand, requires the presence of a catalyst in addition to elevated temperature and pressure.

Benzene as a molecule is used as a synthetic precursor for other benzene-containing molecules. It is used in the chemical industry as a solvent and to make detergents, dyes, lubricants, drugs, etc. Benzene is carcinogenic and should be handled with extreme caution.

Some familiar benzene-containing molecules are trinitrotoluene (TNT), used chiefly as an explosive; naphthalene, used in mothballs; and cumene (also known as isopropyl benzene)—Figure 8.2.

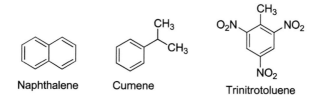

| Naphthalene | Cumene | Trinitrotoluene |

**FIGURE 8.2**   Aromatic compounds.

The structures above are said to be skeletal. Skeletal structures come in handy for molecules that are fairly large since including lines for bonds and writing the carbons, hydrogens, and any other atoms in a structure can be rather messy. Structural presentation of organic compounds is discussed in detail in the next section, Section 8.4.

## 8.3.5   CYCLIC NONAROMATIC HYDROCARBONS

Synthetic and natural carbon-containing compounds that are cyclic (and not aromatic in nature) also exist. Cycloalkanes, for example, are hydrocarbons containing C–C single bonds with a circular connectivity. Cycloalkanes have the general formula $C_nH_{2n}$.

Cycloalkenes are hydrocarbons containing C=C bonds arranged in rings. Cycloalkenes have the general formula $C_nH_{2n-2}$ (Figure 8.3).

cyclopentane        cyclohexene

**FIGURE 8.3**   Cyclic compounds.

Fused cyclic compounds, such as testosterone (male sex hormone) shown in Figure 8.4, play vital roles in biological systems.

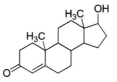

**FIGURE 8.4**   Testosterone.

## 8.4  STRUCTURAL REPRESENTATION OF ORGANIC COMPOUNDS

Organic molecules are represented using condensed structures, Kekulè structures, skeletal (line) structures, space-filling models, or ball and stick models (Table 8.4).

In skeletal structures (also known as line notation), lines are used to denote shared electrons between carbon atoms. One shared electron pair is denoted using a single line (single bond). Two lines (double bond) represent two pairs of shared electrons ($C=C$), and three lines (triple bond) represent three pairs of shared electrons ($C\equiv C$).

Each vertex represents a carbon atom attached to the number of hydrogen atoms required for the carbon to satisfy the octet requirement (four bonds). If more than two carbons are present in a row, then the line is drawn in a zigzag fashion, with each angle break in the line indicating the presence of carbon. Atoms other than carbons and hydrogens attached to carbons must be shown in skeletal structures.

Note that there are two possible structures for the alkane with the molecular formula $C_4H_{10}$ (Table 8.4). Compounds with the same molecular formula but different connectivity of atoms are known as **constitutional isomers**.

Typically, the number of constitutional isomers possible increases with increase in number of carbon atoms. $C_5H_{12}$, for example, has the three isomers shown in Figure 8.5. The three structures have different physical and chemical properties. Isomerism is a major contributor to the vast number of organic molecules in existence.

## TABLE 8.4
### Different Ways of Writing Molecular Structures

| Name | Condensed Structure | Kekulè Structure | Skeletal Structure |
|---|---|---|---|
| Butane ($C_4H_{10}$) | $CH_3CH_2CH_2CH_3$ | | |
| 2-Methyl-propane($C_4H_{10}$) | $CH_3CH(CH_3)CH_3$ | | |
| 1-Propanol ($C_3H_8O$) | $CH_3CH_2CH_2OH$ | | |
| Acetone (2-propanone) | $CH_3COCH_3$ | | |
| 2-Chloro-1-butene | $CH_2C(Cl)CH_2CH_3$ | | |

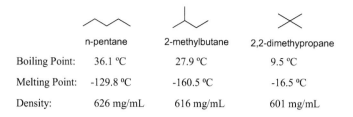

|              | n-pentane   | 2-methylbutane | 2,2-dimethypropane |
|--------------|-------------|----------------|--------------------|
| Boiling Point: | 36.1 °C   | 27.9 °C        | 9.5 °C             |
| Melting Point: | -129.8 °C | -160.5 °C      | -16.5 °C           |
| Density:     | 626 mg/mL   | 616 mg/mL      | 601 mg/mL          |

**FIGURE 8.5**   Constitutional isomers of $C_5H_{12}$.

## 8.5   NAMING ORGANIC COMPOUNDS

Chemists have a well-defined set of rules, known as the IUPAC system of nomenclature (established by the International Union of Pure and Applied Chemistry in 1961)
for naming organic compounds. Most of these names stem from alkane names.

Compound names are normally listed in scientific literature systematically and
thus require the reader to be familiar with the IUPAC naming system. Tables 8.1–8.3
provide the names of the first few aliphatic hydrocarbons. It is important to realize
that they only differ in the ending: -ane for alkanes; -ene for alkenes; and -yne for
alkynes. It is worthy to note that while, on occasion, there may be more than one
"common" name for an organic compound, there is usually only one correct "chemical" name, according to the IUPAC system. An example of this is acetone and
dimethyl ketone as common names for the chemically correct name of 2-propanone.

Numbering of carbon atoms is done during the naming process to give double
bonds and triple bonds the lowest possible numbers. Substituents should also get
lower numberings, with multiple substituents listed alphabetically.

Tables 8.5 and 8.6 detail the names and skeletal structures of some alkenes and
alkynes, respectively. (See Table 8.4 and Figure 8.5 for sample alkane names.)

---

**TABLE 8.5**
**Alkenes**

| Molecular Formula | Name | Skeletal Structure |
|-------------------|------|--------------------|
| $C_5H_{10}$ | 1-Pentene | |
|  | 2-Methy-1-butene | |
|  | 3-Methyl-1-butene | |
|  | 2-Methyl-2-butene | |
|  | *cis*-**2-Pentene**[a] | |
|  | *trans*-**2-Pentene**[a] | |

[a]  Geometric Isomers (see Section 8.6).

---

**TABLE 8.6**
**Alkynes**

| Molecular Formula | Name | Skeletal Structure |
|---|---|---|
| $C_6H_{10}$ | 1-Hexyne | |
| | 2-Hexyne | |
| | 3-Hexyne | |
| | 3-Methyl-1-pentyne | |
| | 4-Methyl-1-pentyne | |
| | 4-Methyl-2-pentyne | |
| | 3,3-Dimethyl-1-butyne | |

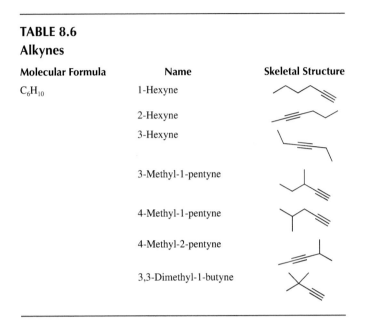

**Example 8.1:** (a) & (b) Name Two Hydrocarbons From Their Molecular Structures

Systematically, name the compounds shown below.

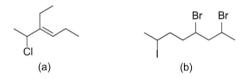

(a)                    (b)

*Solution*

First, identify the longest C–C chain with the most substituents (if any). Next, identify the functional groups and their locations. Number the molecule along the longest C–C chain starting from the end that will give the first substituent(s) the lowest possible number. Finally, assemble the name.

a. Longest chain has six carbons with a double bond on the third carbon; hence, 3-hexene is the parent name. Number from left to right as that places the first substituent (chloro-) on the second carbon and the second substituent (ethyl)) on the third carbon. (This is lower than if you were to number from right to left as that would place the ethyl- substituent on fourth carbon and the chloro- substituent on the fifth carbon.)

Compound (a) is 2-chloro-3-ethyl-3-hexene

b. Longest chain has eight carbons; hence, octane is the parent name. Number from right to left as that locates the substituents on carbons 2, 4, and 7. (This is lower than if you were to number from left to right as that would place substituents on carbons 2, 5, and 7.) Use prefixes di-, tri-, etc., for multiple substituents of the same kind.

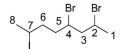

Compound (b) is 2,4-dibromo-7-iodooctane.

## 8.6  ISOMERS AND CONGENERS

There are a total of six structures for the molecule $C_5H_{10}$ (Table 8.5). The first four differ in connectivity (arrangement of atoms). They are **constitutional isomers**.

The last two, ***cis*-2-pentene** and ***trans*-2-pentene** (Table 8.5 and Figure 8.6), have the same connectivity but different spatial orientation. They are **geometric stereoisomers**—molecules with the same molecular formula and connectivity but different orientation of atoms/groups of atoms in space. *Cis*- denotes that the higher-priority substituents (substituents with higher atomic number) are on the same side of the double bond while *trans*- denotes that the higher-priority substituents are on opposite sides.

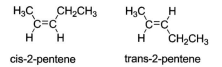

**FIGURE 8.6**  *cis*-2-Pentene and *trans*-2-pentene.

Thus, **isomers** are molecules with the same molecular formula but different structural formulas. There are two types of isomers: **constitutional isomers** and **stereoisomers**. Constitutional isomers differ in connectivity while stereoisomers have the same connectivity but different orientation in space.

There are two types of **stereoisomers**, geometric (***cis-trans***) isomers and **stereoisomers** with one or multiple **asymmetric centers**. In the next chapter, the concept of stereoisomers with asymmetric centers (also known as stereocenters or chirality centers) will be discussed in more detail.

**Congeners**: Organic molecules can further be grouped based on number of similar substituents. Congeners are molecules that have the same carbon skeleton but differ by the number of same substituents and/or substitution pattern. For example, 2-chlorobiphenyl, 2,4'-dichlorobiphenyl and 2,2', 3-trichlorobiphenyl (Figure 8.7) are polychlorinated biphenyl (PCB) congeners.

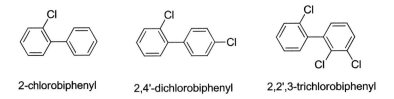

**FIGURE 8.7** Polychlorinated biphenyl (PCB) congeners.

The number of chlorine atoms and their location in a PCB molecule determine its physical and chemical properties. PCBs are known for their chemical stability, high boiling point, nonflammability, and electrical insulating properties. PCB synthesis was banned in the United States in 1979 due their range of toxicity and nonbiodegradability. Although no longer commercially produced in the United States, PCBs may be present in products and materials produced before 1979 such as transformers and capacitors, electrical equipment such as voltage regulators and switches, oil-based paint, thermal insulation material including fiberglass, foam, cork, etc.

## 8.7 BIOFUELS

Global climate change is driven by the use of fossil fuels. Fossil fuels were formed over millions of years (about 50–300 million years) from the bodies of dead algae, plant, and some animal life buried beneath the surface of the earth. The problem is that all of the carbon trapped by this process over a very long period of time is now being released very rapidly. Nonetheless, fuels are needed to run cars, heat homes, and make electricity. One answer to this problem is to use biofuels.

Biofuels tend to be chemically very similar to fossil fuels, but are made from plants that are produced today. This is a very important distinction. Living organisms are composed predominantly of carbon, and for plant life, the carbon they use to build themselves comes from the air in the form of $CO_2$. This process is referred to as "fixing" carbon or harvesting carbon to form the molecules that make up the plant. In this process, $O_2$ is released. When fuel is burned, the carbon in the plants is converted back to $CO_2$ and released into the atmosphere. This cycle is said to be carbon neutral. The plants are built up of $CO_2$ to make carbon compounds, and the same amount of $CO_2$ is released when the plant-based fuel is burned. Thus, the $CO_2$ released is then recaptured by the next crop of plants to be turned into fuel (Figure 8.8).

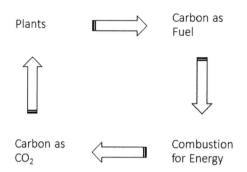

**FIGURE 8.8**  Carbon-neutral biofuel cycle.

The most common biofuel today is ethanol. Ethanol fuel does require some engine modifications, as some materials used in seals and gaskets can be corroded by ethanol, but the cost impact is minimal. Ethanol can be made readily from sugar in plant life by fermentation, the same process used to make beer or wine.

$$C_{12}H_{22}O_{11} \rightarrow 2\ CO_2 + 2\ C_2H_5OH$$

The ethanol can then be burned as a fuel.

$$C_2H_5OH + 3\ O_2 \rightarrow 2\ CO_2 + 3\ H_2O + energy$$

In the United States, ethanol is now used as an additive in gasoline to displace some of the fossil fuel with renewable biofuel by government mandate. In Brazil, high ethanol content (20%–25%) is used in almost all automobiles as fuel. Brazil has an ideal climate for growing sugar cane, which allows for this very high-volume production of ethanol for automotive use.

Another common biofuel is biodiesel. Biodiesel is a somewhat ambiguous term, though it is meant to refer to FAME or Fatty Acid Methyl Ester, a chemically converted molecule derived from vegetable oil. Vegetable oil is also sometimes referred to as biodiesel, as diesel engines can be run on vegetable oil, but is not what is technically meant by the term biodiesel. Biodiesel is produced by the chemical action of sodium or potassium methoxide on vegetable oil as shown below. The process is an equilibrium process, but it favors the products side heavily (Figure 8.9).

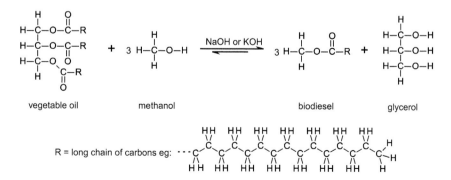

**FIGURE 8.9**  Chemical reaction scheme for synthesis of biodiesel.

Biodiesel can be used as a substitute for diesel with zero modification to the engine. It is most commonly used as a blend, as most car manufacturers do not permit use of more than 20% biodiesel content fuel under their warranties. Biodiesel, however, tends to burn very cleanly and even dissolves grease deposits in engines and fuel lines, leading to cleaner engines.

The two main problems with biofuels are cost and total availability. Fossil fuels are generally still currently cheaper than biofuels. Most biofuels access the market either through government subsidy or by government mandate. As long as fossil fuels are readily and cheaply available, biofuels cannot compete without political support.

Even with governmental support, one place where biofuels run into a problem is that biomass is required for their production. Some sources, such as corn, are very poor, as corn has relatively little accessible convertible sugars present. Sugar cane, or its engineered offshoot energy cane, has a much higher sugar content, but requires very specific climates for growth. This limits the availability of raw materials for production of fossil fuels.

A solution for the shortage of raw material is to convert cellulose, the structural compound in plants, into liquid fuel as well. This, however, is chemically difficult, as the process for converting cellulose into smaller molecules is much more demanding energetically and cost wise. Cellulosic ethanol research, the effort to produce ethanol from cellulose rather than starch or sugar, is ongoing.

Biofuels, as a carbon neutral fuel source, can be a key component in mitigating carbon emissions. Governmental incentives, as well as development of improved technologies for producing biofuels from currently unusable raw materials will, however, be necessary for the success of biofuels.

# 9 Asymmetric Centers, Functional Groups, and Characterization

## 9.1 ASYMMETRIC CENTERS: ENANTIOMERS AND DIASTEREOMERS

**Enantiomers** are a pair of stereoisomers that are nonsuperimposable mirror images. A molecule with one asymmetric center will exist in two enantiomeric forms. An asymmetric center is an atom in a molecule that is bonded to four different atoms/groups of atoms. Enantiomeric molecules have no plane of symmetry and are said to be **chiral** (Figure 9.1). Conversely, molecules with a plane of symmetry are said to be **achiral**. Achiral molecules are superimposable onto their mirror images (Figure 9.2).

chirality/asymmetric center

**FIGURE 9.1** Asymmetric center in 2-butanol (chiral molecule).

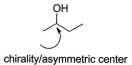

Bromodichloromethane

**FIGURE 9.2** An achiral molecule.

Molecule bisects along H–C–Br plane, one-half of the molecule is mirror image of the other half.

**Chirality** is similar to "handedness"—the reason your left-hand glove will not fit in your right hand is because your two hands are nonsuperimposable mirror images. Your hand has no plane of symmetry and is said to be chiral. Your right and left hands constitute a pair of enantiomers. Figure 9.3 shows the two enantiomers of 2-butanol.

DOI: 10.1201/9781003092759-9

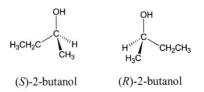

(S)-2-butanol        (R)-2-butanol

**FIGURE 9.3**    Enantiomers of 2-butanol.

Enantiomers have similar physical and chemical properties in achiral environments, but different properties in chiral environments. To help understand this, consider an oven mitten—an oven mitten is achiral—either hand, left or right, can fit in. A pair of gloves, on the other hand, are chiral—right hand can only, comfortably, fit in the right-hand glove and left hand can only, comfortably, fit in the left-hand glove.

One distinguishing characteristic of enantiomeric pairs is their ability to rotate plane-polarized light to the same magnitude but in opposite directions.

**Diastereomers** are stereoisomers that are not mirror images. They occur in molecules with multiple asymmetric centers. Diastereomers have different chemical and physical properties and can be easily separated from one another using the usual separation techniques—column chromatography, distillation, fraction crystallization, etc. (Figure 9.4).

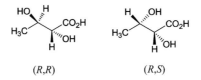

(R,R)                (R,S)

**FIGURE 9.4**    Diastereomers.

## 9.2   SEPARATION OF ENANTIOMERS

The ability of enantiomers to behave differently under chiral conditions enables their separation. The process of separating enantiomers is known as enantiomeric resolution. A racemic mixture may require separation if one enantiomer diminishes the activity of the other, or if one enantiomer is effective in producing the desired therapeutic results while the other has adverse side effects.

Enantiomeric pairs are separated by reacting the mixture with a chiral (enantiomerically pure) compound, thereby converting the two into diastereomers. Diastereomers have different chemical and physical properties and can thus be separated using the usual separation techniques. This separation is then followed by a chemical transformation to cleave off the attached chiral reagent from the two diastereomers.

Enantiomers can also be separated using column chromatography equipped with chiral material. The two enantiomers reversibly bind to the chiral material at different rates, thereby enabling their separation. Enantiomers with dramatic pharmacological activity differences often require that they are separated before being marketed. It is important to note that separation of enantiomers can be tedious and expensive, and some drugs are still marketed as racemic mixtures.

Citalopram (Figure 9.5), for example, is a racemic mixture of R(−)- and S(+)-enantiomers. Citalopram's antidepressant activity has been linked to the S(+)- enantiomer, also known as escitalopram. The presence of R(−)- enantiomer is known to inhibit the pharmacological effect of the S(+)- enantiomer. It is therefore necessary that the two be separated before being marketed.

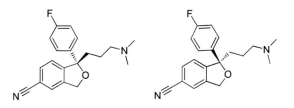

**FIGURE 9.5**    S(−)- and R(+)- citalopram.

Vitamin C, ascorbic acid (Figure 9.6), also exists in two enantiomeric forms with L-ascorbic acid isomer being the one that is biologically active and useful for human consumption.

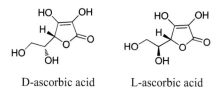

D-ascorbic acid        L-ascorbic acid

**FIGURE 9.6**    D- and L-ascorbic acid.

Similarly, Vitamin E or the tocopherol family has eight stereoisomers, but only the alpha-tocopherol (Figure 9.7) is biologically active and useful for human consumption.

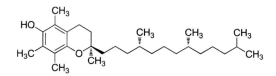

**FIGURE 9.7**    Alpha-tocopherol.

The difference in pharmacological activity of an enantiomeric pair can be attributed to the chiral nature of human body. (Remember enantiomers show different chemical and physical properties under chiral conditions.) We can liken the bodily "discomfort" with an incompatible enantiomer to the left hand in right-hand glove scenario.

It is worth noting here that while scent seems to depend largely on the shape and structural makeup of a molecule, stereoisomerism plays a role in some cases. For example, carvone exists as an enantiomeric pair, *R*-(−)- and *S*-(+)-carvone (Figure 9.8). *R*-(−)-carvone has a sweetish, minty smell like spearmint leaves while *S*-(+)-carvone has a spicy aroma like caraway seeds. The fact that two enantiomers can have different scents is proof that olfactory receptors do contain chiral groups.

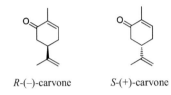

*R*-(−)-carvone     *S*-(+)-carvone

**FIGURE 9.8**   *R*-(−)- and *S*-(+)-carvone.

Similarly, *R*-(+)-limonene has a fresh citrus, orange-like scent, while *S*-(−)-limonene has harsh, turpentine-like scent (Figure 9.9).

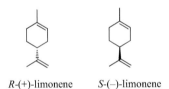

*R*-(+)-limonene     *S*-(−)-limonene

**FIGURE 9.9**   *R*-(+)- and *S*-(−)-limonene.

It is important to note that scent differences in stereoisomers do not always apply. It is also interesting to note two different molecules with very different structures, shapes, or functional groups, may give nearly the same scent. This is true of the musk scent found in men's colognes, where four very different molecules give nearly the same scent. [See McFadden and Khalili, 2014 for more details and actual structures.]

## 9.3   FURTHER CLASSIFICATION OF ORGANIC COMPOUNDS—FUNCTIONAL GROUPS

Organic compounds are further classified into families based on the **functional groups** they contain. A **functional group** is the group of atoms in a molecule that determines the molecule's reactivity. Double and triple bonds are the reactive sites in alkenes and alkynes, respectively, and are therefore said to be their functional groups. Table 9.1 details common functional groups in organic chemistry.

## TABLE 9.1
## Functional Groups

| Family | General Structure | Functional Group | Example |
|---|---|---|---|
| Alcohols | R–ÖH | –ÖH | ∧OH<br>ethanol |
| Alkenes | (H)R$\quad$R(H)<br>$\quad$C=C<br>(H)R$\quad$R(H) | = | ⤳<br>propene |
| Alkynes | (H)R≡R(H) | ≡ | H–≡–H<br>acetylene |
| Amines | R–NH$_2$ | –NH$_2$ | CH$_3$CH$_2$NH$_2$<br>ethylamine |
| Ethers | R–Ö–R | –Ö– | ∧O∧<br>diethyl ether |
| Aldehydes | :O:<br>R⌢H | :O:<br>⌢H | :O:<br>H⌢H<br>formaldehyde |
| Ketones | :O:<br>R⌢R | :O: | :O:<br>H$_3$C⌢CH$_3$<br>acetone |
| Carboxylic acids | :O:<br>(H)R⌢Ö–H | :O:<br>Ö–H | :O:<br>H$_3$C⌢OH<br>acetic acid |
| Esters | :O:<br>(H)R⌢Ö–R' | :O:<br>Ö– | :O:<br>H$_3$C⌢OCH$_2$CH$_3$<br>ethyl acetate |
| Amides | :O:$\quad$R"<br>(H)R⌢N:<br>$\qquad$R' | :O:<br>⌢N: | :O:$\quad$H<br>H⌢N<br>$\qquad$H<br>formamide |
| Thiols | R–ṠH | –ṠH | ∧SH<br>ethanethiol |

**Example 9.1:** Advanced Nomenclature—Name Five Organic Compounds With Different Functional Groups (a), (b), (c), (d), and (e)

Name the following organic molecules:

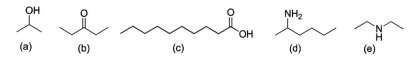

(a)$\qquad$(b)$\qquad$(c)$\qquad$(d)$\qquad$(e)

Answers: (a) 2-propanol (common name: isopropyl alcohol) (b) 3-pentanone, (c) decanoic acid, (d) 2-aminohexane, (e) diethylamine.

**Example 9.2:** Draw Structures for Four Organic Compounds With Different Functional Groups

Provide structures for the following:
    (a) 1,2-ethanediol, (b) ethylmethyl ether (or methoxyethane), (c) 2-propanone (common name: acetone), (d) methanal (common name: formaldehyde).
Answers:

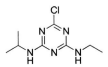

    (a)         (b)         (c)         (d)

Some organic compounds, such as atrazine (Figure 9.10), may have more than one functional group per molecule. Atrazine is a widely used herbicide that just got reapproved for use by the U.S. EPA but on a more limited basis after a 7-year review study due to its developmental and reproductive toxicity; currently banned for use in the EU. This issue is discussed in greater detail in Chapter 15.

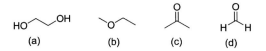

**FIGURE 9.10** Common name: Atrazine. IUPAC name: 1-Chloro-3-ethylamino-5-isopropylamino-2,4,6-triazine.

    Similarly, theobromine (Figure 9.11) is a bitter alkaloid of the cacao plant, with the chemical formula C7H8N4O2, found in chocolate.

**FIGURE 9.11**    Theobromine.

## 9.4    SOME COMMON PRODUCTS AND FUNCTIONAL GROUPS

**Methanol (CH$_3$OH)**
    Functional group: C–OH
    Used as an organic solvent.
    Highly toxic. Can cause blindness and/or death upon inhalation or ingestion.

### Ethanol (CH₃CH₂OH)

Functional group: C–OH

Main component of alcoholic beverages. Used as a motor fuel.

Irresponsible consumption can cause alcohol poisoning and/or other health-related issues.

### Ethylene glycol (HOCH₂CH₂OH)

Functional group: Two C–OH groups.

Used as antifreeze, in printing inks, paint solvents, and hydraulic fluids.

**Cholesterol**

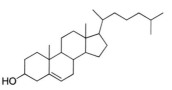

Functional group: C–OH, C=C

A naturally occurring source of alcohol found in most animal tissues and egg yolks.

**Retinol (Vitamin A alcohol)**

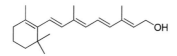

Functional group: C–OH, C=C

A naturally occurring source of alcohol extracted from fish liver oils.

**Isoamyl (or isopentyl) acetate (IUPAC name: 3-methybutyl ethanoate)**

Functional group: Ester

Responsible for the banana flavor.

**Ethyl hexanoate**

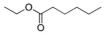

Functional group: Ester.

Responsible for the pineapple flavor.

Note that some naturally occurring products exist as mixtures. For example, coffee is a mixture of over 200 compounds, and as such, regardless of specific blends and tastes, its flavor has never been successfully reproduced synthetically.

## 9.5   STRUCTURAL ANALYSIS

Knowing a detailed structure of a molecule is crucial to understanding the properties and reactivity of the molecule. Structural knowledge is also necessary for synthetic reasons. Naturally occurring molecules with medicinal potential, for example, often need to be synthesized to ensure a continuous supply and worldwide marketing. Below are the common analytical methods used by chemists for structural elucidation. In most cases, a compound (isolated in nature or synthesized in the lab) is subjected to all the first three techniques discussed below for its definitive identification.

### 9.5.1   NUCLEAR MAGNETIC RESONANCE (OR NMR) SPECTROSCOPY

This technique is mostly used to detail C–C and C–H connectivity in a molecule. It gives the total number of carbon and hydrogen atoms and can be used to determine the point of attachment of a hydrogen or carbon. It can be used to determine the different kinds of carbons (primary, secondary tertiary, and quaternary) and hydrogens a compound has. NMR spectra can also be obtained for N, F, and P nuclei. Nuclear magnetic resonance technique (modified to magnetic resonance imaging, MRI) is medically used to examine internal organs and structures. Although expensive, it is noninvasive and safer than the ionizing radiation of X-rays.

### 9.5.2   INFRARED (OR IR) SPECTROSCOPY

This technique is mainly used to confirm the presence or absence of a given functional group (O–H, $C=O$, $C\equiv C$, N–H, etc.) in a molecule. The shape, the size, and the location of an absorption band help identify the possible functional groups in a molecule. The absence of bands at specific wavenumbers indicates the absence of the functional group in a molecule. The intensity of absorption depends on bond polarity, whereas absorption location on the spectrum depends on bond strength. In pharma industry/medical field, IR is used to investigate structural, functional, and compositional changes in biomolecules, cells, and tissues. IR spectroscopic techniques can, for example, be used to identify molecular changes resulting from obesity and the rightful drugs/doses for treatment. IR technique is known for its rapid results, accuracy, noninvasiveness, and cost-effectiveness.

### 9.5.3   MASS SPECTROSCOPY (OR MS)

This technique is mainly used to determine molecular weights and molecular formulas of molecules through molecular ionization and fragmentation. Each molecule has its own unique fragmentation pattern; as such mass spectrum data is like a fingerprint of a molecule. Structural information for a given molecule can be obtained from the fragments. MS is used in drug testing and discovery, forensic toxicology, carbon dating, protein identification and quantification, pesticide residue analysis, etc. Mass spectrometry is quick, cost-effective, and accurate.

### 9.5.4    GAS CHROMATOGRAPHY–MASS SPECTROSCOPY (OR GC–MS) AND HIGH-PERFORMANCE LIQUID CHROMATOGRAPHY (OR HPLC)

MS coupled with gas chromatography (GC) or liquid chromatography (LC) system is commonly used for definitive identification of compounds with low detection limits and the potential for quantitative analysis. The collected sample mixture is first separated using chromatography and then sequentially passed through the spectrometer for ionization, separation, and identification. While GC-MS is great for volatile compounds, HPLC is applicable to a greater variety of mixtures including nonvolatile or thermally unstable molecules. GC-MS and HPLC are advantageous for their two-fold information for each analysis—retention time and mass spectral information. Both techniques can be used to obtain evidence in situations involving drugs and fire debris, test residual solvents, identify trace impurities in solvents or gases, evaluate extracts from plastics, etc. They are highly versatile, sensitive, and can be used to analyze complex mixtures.

## 9.6    ORGANIC SOLVENTS

Organic solvents are routinely encountered in industrial and commercial processes and in product formulations. Their molecular structures determine their different physical properties and hence specific applications. Five of them are compared in Table 9.2. For a more complete list of solvents and their properties, the interested reader is encouraged to consult *The Handbook of Organic Solvents* by David R. Lide, CRC Press, 1995.

**TABLE 9.2**
**Common Organic Solvents and Their Properties**

| Name | CASRN | Melting Point °C | Boiling Point °C | Density at 25°C | Solubility Limits | Toxicity as TLV/TWA |
|------|-------|------------------|------------------|-----------------|-------------------|---------------------|
| Acetone or 2-Propanone [(CH$_3$)$_2$CO] | 67-64-1 | −94.8 | 56.0 | 0.7899 g/mL | V.Sol. in H$_2$O, Benz, EtOH | 750 ppm |
| Isopropanol or 2-Propanol [(CH$_3$)$_2$CHOH] | 67-63-0 | −89.5 | 82.3 | 0.7855 g/mL | V.Sol. in H$_2$O, EtOH, Benz | 400 ppm |
| Benzene [C$_6$H$_6$] | 71-43-2 | 5.5 | 80.0 | 0.8765 g/mL | V.Sol. in EtOH, Ether, Acetone | 10 ppm |
| Hexane [C$_6$H$_{14}$] | 110-54-3 | −95.3 | 68.7 | 1.3749 g/mL | V. Sol. in EtOH | 50 ppm |
| Ethylene Glycol [HOCH$_2$CH$_2$OH] | 107-21-1 | −13 | 197.3 | 1.1088 g/mL | V.Sol. in H$_2$O, Acetone, EtOH | 50 ppm |

# 10 The Essentials of Polymer Chemistry

## 10.1 THE BASICS

Polymers are materials whose molecules are made up of many (poly~) small, repeating subunits (~mers), present initially as single units or individual molecules called monomers. Monomers undergo a reaction to form polymers. The chief reason this happens and produces a dizzying array of carbon-based polymers is that one of carbon's properties (more mentioned in Chapter 8) is its unique ability to bond to itself. The most familiar class of these polymers is plastics, which are a broad variety of different, synthetic compounds. Many natural materials, such as rubber and cellulose (the main structural component of wood), however, are also polymers.

A couple of definitions are important when talking about polymer chemistry. The precursor compound for a polymer is referred to as a monomer. Monomers are the individual molecules that react to form the compound. Segmers are what monomers turn into once the polymer is formed. They are repeating segments of the polymer. Oligomers are polymers that only have a smaller number of individual subunits, typically on the order of tens of segmers.

## 10.2 ADDITION POLYMERS

Addition polymers are a class of polymers where bonds between doubly bonded carbons are formed in order to link together individual subunits. The consequence is either the loss or movement of the double bonds. The simplest case is that of polyethylene (PE), a very common plastic. PE is a polymer of ethene (ethylene is a common or trivial name for ethene). Figure 10.1 depicts this transformation. Arrows indicate the movement of one of the bonds from a double bond into a new bond between the monomers.

Here the double bond of an ethene molecule is reduced to a single bond, as it makes a connection to the next ethene molecule in line. This causes a chain reaction, as the new bond forces the next ethene molecule to lose its double bond and form a single bond to a third ethene. This chain reaction propagates until thousands of units of ethenes are bound into a chain.

**FIGURE 10.1** Addition polymerization.

DOI: 10.1201/9781003092759-10

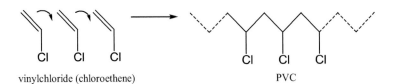

**FIGURE 10.2**   Formation of polyvinylchloride (PVC).

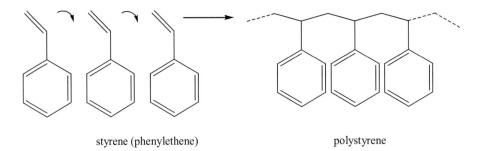

**FIGURE 10.3**   Formation of polystyrene.

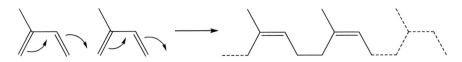

**FIGURE 10.4**   Complex addition polymerization in the formation of rubber.

Other commonly encountered plastics, such as polyvinylchloride and polystyrene (PS), are also formed by additional polymerizations. Both of these plastics have a subunit that can be seen as an ethene with a substituent replacing one of its hydrogens. For polyvinyl chloride (PVC), the substituent is chlorine, for PS, the substituent is a benzene ring (the technical term for this is a phenyl substitution). This concept is illustrated in Figures 10.2 and 10.3.

Rubbers are also formed from an addition polymerization, which is slightly more complex. The basic subunit here is an isoprene, which has two double bonds. The isoprene loses the double bonds at the two ends of the molecule, but gains one in the middle, for a net loss of one double bond, and the formation of a polymer chain as shown in Figure 10.4.

## 10.3   CONDENSATION POLYMERS

Another class of polymers are condensation polymers. As the name suggests, this type of polymerization generates water, or some other condensate, as a by-product. Two common polymers formed by condensation are nylon, of which there are many varieties, and polycarbonates.

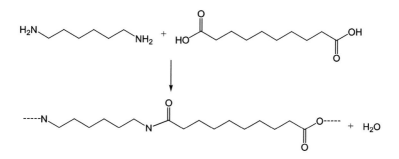

**FIGURE 10.5** Condensation polymerization of nylon.

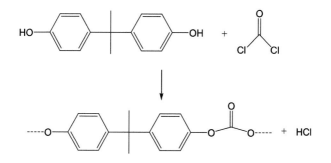

**FIGURE 10.6** Condensation polymerization to form polycarbonates.

Nylon is formed from the reaction of a diamine and a dicarboxylic acid. The length of the carbon chain between the two amines and the two carboxylic acids will govern the mechanical properties of the polymer produced. Nylon is usually named as nylon-$x$, $y$; here, $x$ is the number of carbon atoms in the diamine compound, and $y$ is the number of carbon atoms in the dicarboxylic acid. The schematic for the synthesis of nylon 6,10 is shown in Figure 10.5.

Another important commodity chemical is polycarbonate. Polycarbonates are plastics that are hard, and also clear, and thus can be used as a substitute for glass in certain applications. The most common polycarbonate is made from bisphenol A and phosgene. The reaction shown in Figure 10.6 produces HCl as a condensate instead of $H_2O$, but the principle is the same.

## 10.4 COMMON POLYMERS

Table 10.1 lists some common polymers and their structures. They are also identified by their recycling code, which can often be found on plastic products.

Of note are high-density polyethylene (HDPE) and low-density polyethylene (LDPE), which are very common plastics coded, differently, but chemically apparently identical. HDPE and LDPE are made of the same monomer, but their larger-scale structures are different. HDPE has very straight chains of PE, and these stack on each other neatly to make a harder, denser, higher-melting-point plastic. PE in

**TABLE 10.1**
**Common Polymers**

| Name | Code | Structure | Example Uses |
|---|---|---|---|
| Polyethylene terephthalate (PETE or PET) | [1] PETE | | Soda and water bottles, sails, peanut butter jars |
| High-density polyethylene (HDPE) | [2] HDPE | | Grocery bags, milk jugs, plastic lumber, detergent and motor oil bottles, butter tubs |
| Polyvinyl chloride (PVC or V) | [3] V | | Pipes, shower curtains, cleaning agent bottles |
| Low-density polyethylene (LDPE) | [4] LDPE | | Plastic bags, 6 pack rings, tubing |
| Polypropylene (PP) | [5] PP | | Ropes, packaging materials, food containers, syrup bottles, bottle caps, plastic diapers |
| Polystyrene (PS) | [6] PS | | Insulation, plastic utensils, Styrofoam cups |
| Polytetrafluoroethylene (Teflon) | * [7] OTHER | | Cookware, low friction surfaces |
| Polyamide (Kevlar) | * [7] OTHER | | Body armor |

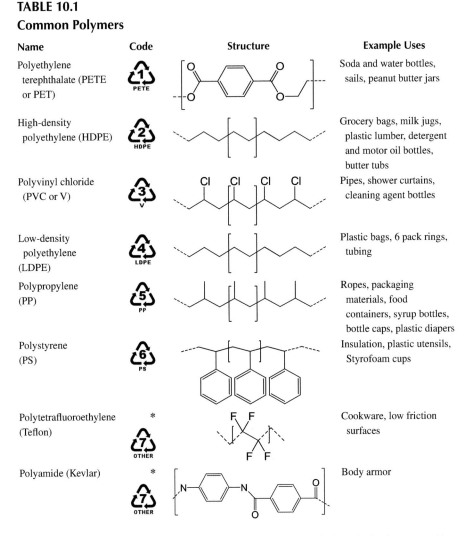

Brackets denote the segmers of the shown polymer. *Code 7 classification is for "other" polymers, anything that does not fall into the other six categories. Some examples of other polymers are given.

LDPE tends to have branches, and this leads to the strands stacking relatively poorly. This makes LDPE softer, less dense, and lower melting relative to HDPE.

## 10.5   POLYMER TYPES

Once formed, plastics generally fall into one of two broad categories: thermoplastics and thermosets. Thermoplastics are more common than thermosets. The principle difference is the ability to undergo numerous melt and solidification cycles without

---

**TABLE 10.2**
**Examples of Thermoplastic and Thermoset Plastics**

| Thermoplastic Plastics | Thermoset Plastics |
|---|---|
| High-density polyethylene (HDPE) | Epoxy |
| Polyvinyl chloride (PVC) | Cyanate ester |
| Polypropylene (PP) | Polyester |
| Polycarbonate (PC) | Polyurethane |
| Polyethylene terephthalate (PET) | Silicone |
| Acrylonitrile butadiene styrene (ABS) | Vulcanized rubber |
| Polyamide (PA) | Bakelite |

---

significant degradation in structural integrity. Thermoplastics are usually supplied in the form of small pellets or sheets, which are heated and formed into the desired shape using one of eight manufacturing processes, e.g., blow molding, injection molding, polymer casting, and extrusion. The process is usually reversible since no chemical bonding takes place, making recycling (or melting) and reusing thermoplastics feasible.

In contrast, thermosetting plastics, or simply thermosets, remain in a permanent solid state after curing. Polymers in thermoset materials undergo a cross-linking process that is induced by heat, light, or suitable radiation. The curing or cross-linking process forms irreversible chemical bonds. Thus, thermoset plastics decompose upon heating and will not reform upon cooling. The first synthetic plastic, Bakelite, was made from the condensation reaction between phenol and formaldehyde in 1907 and is a thermoset plastic. Recycling thermosets is very difficult if not impossible with current technologies. Table 10.2 lists examples of common thermoplastics and thermosets.

## 10.6 POLYMER PROBLEMS

A major concern surrounding polymers is the potential for environmental contamination from use of plastics. This can occur from leeching of unreacted monomers from the plastic matrix into the environment, from the breakdown of the plastic from years of exposure to UV light from the sun, or from mechanical degradation due to frictional forces from constant use.

For example, a major controversy has surrounded the use of polycarbonate plastics in beverage bottles. When a polymerization reaction on the industrial scale occurs, not all of the monomers react; some remain trapped in the plastic matrix, unreacted. One of the monomers, bisphenol A, has been shown to have a biological effect, as it mimics the functions of certain hormones in the body. It is an estrogen mimic and an endocrine-disrupting chemical (EDC). EDCs are discussed in more detail in Chapter 15. This can potentially affect the long-term health of people who use the water bottles, as, over time, small amounts of the bisphenol A would leech into the water and would end up being ingested by the owner. Alternative plastics with different feedstocks were eventually developed to bypass this problem.

PVC faces similar potential concerns. Vinyl chloride, the monomer from which polyvinylchloride is made, is a known human carcinogen. Unreacted monomer, which may be trapped in the PVC matrix, can leech out into the atmosphere, especially if the PVC is exposed to heat or sunlight, and be subsequently inhaled or absorbed through the skin. Furthermore, if PVC is burned, it releases corrosive HCl gas. While PVC is a hard solid at room temperature, it can be made pliable and flexible with the addition of one or more chemical substances known as plasticizers. Common examples include di-octyl-phthalate (DOP) and di-2-ethylhexyl phthalate (DEHP), both of which are carcinogenic (discussed in more detail in Chapter 15). Garden hoses, upholstery and flooring material, shower curtains, and cable coatings are just some of the applications of plasticized PVC. But as PVC ages and is exposed to heat or sunlight, these plasticizers migrate out of the plastic matrix, leaving behind a more brittle solid, subject to slow degradation and disintegration. Thus, PVC has the potential to release not only toxic, unreacted monomers and plasticizers but also microplastics (discussed in more detail in Section 10.8) into biosphere. These concerns have led to the implementation of health and safety regulations, and have to be carefully considered when using or disposing of PVC.

On the other hand, many plastics are polymers that are quite stable and resilient to natural decomposition. Unfortunately, this creates another problem, the problem of plastic waste and its disposal and/or recycling, discussed in more detail in the following section.

## 10.7   PLASTIC WASTE AND THE RECYCLING OF PLASTICS

According to an article in the July 2020 issue of Consumer Reports, 76% of plastic waste ends up in landfills, 16% is incinerated, and 1% litters the oceans, implying that less than 10% actually gets recycled. A more recent issue of Consumer Reports, March 2021, claims that only HDPE and polyethylene terephthalate (PET) get recycled, constituting a total of about 25% of all plastic waste. The exact amount that gets recycled, by total or by category, seems to vary by source of information and is a fluid figure. In another article in Chemical and Engineering News (C&EN) in July 24, 2017 (Geyer et al. 2017), of the 8.3 billion metric tons of virgin plastics that have ever been produced, approximately 60% has been discarded into landfills, 30% has been recycled and is still in use today, and 10% has been incinerated. And in a more recent C&EN article, October 7, 2019, EPA estimates that of the total amount of plastic waste generated each year, only 9% in the United States (and 15% in Europe) actually gets recycled into useful products. According to EPA, the recycled plastic breaks down as follows (Table 10.3).

It is clear from this table that not all plastic waste is recycled and/or reused at the same rate. Some plastics may be reused but not recycled while some may be recycled but not reused. For example, HDPE may be both reused and recycled and has one of the highest recycling/reusing rates. LDPE may be reused but not recycled. PETE may be recycled but not reused. Technology has been developed to recycle PS but is not widely available. It should be noted that not all thermoplastics can or should be recycled. PVC should not be reused or recycled.

**TABLE 10.3**

**U.S. Plastic Waste Generated and Recycled in 2015 by Category**

| Plastic Category | Amount Generated | Per Cent Recycled/Reused |
|---|---|---|
| Low-Density and linear polyethylene (LDPE) | $7.21 \times 10^6$ metric tons | 6.2% |
| Polypropylene (PP) | $7.03 \times 10^6$ metric tons | 0.9% |
| High-density polyethylene (HDPE) | $5.49 \times 10^6$ metric tons | 10.3% |
| Polyethylene terephthalate (PET or PETE) | $4.64 \times 10^6$ metric tons | 18.4% |
| Polystyrene (PS) | $2.15 \times 10^6$ metric tons | 1.3% |
| Polyvinyl chloride (PVC) | $8.09 \times 10^5$ metric tons | <1% |
| Other resins (e.g., nylon, polycarbonate, polyurethane, copolymers, etc.) | $3.98 \times 10^6$ metric tons | 22.6% |

The two chief problems with recycling plastic waste are sorting one category from another and decontaminating each plastic stream from fillers, additives, or leftover food. An additional problem lies in dealing with mixed plastics, i.e., copolymers like ABS plastic, which present their own separate challenge. Global plastic production is projected to quadruple by 2050, so this will be an increasing problem. At present, there are five leading technologies or methods that address the recycling problem. In brief, they are as follows:

I. *Mechanical*: This is mechanical sorting, which starts at the curbside for pickup and is the most common recycling method. Plastics are sorted by category, washed, shredded, and processed into pellets. This method works well for HDPE and PET because they are available in large volume. However, given the degree of contamination by foreign matter and other plastics, as well as degradation over time, mechanically recycled plastics are down-cycled and cannot be used in applications involving food packaging.

II. *Pyrolysis*: This method uses temperatures over 400°C in an oxygen-free process to decompose polymers into smaller hydrocarbons, such as diesel and naptha molecules, which can be subsequently reassembled into new polymers and plastics. One positive point for this method is that it can used to process mixed plastic waste, i.e., copolymers.

III. *Depolymerization*: Another pathway to recycling PET is to subject it to methanolysis and hydrolysis. This process will actually decompose PET into individual monomers of ethylene glycol and dimethyl terephthalate, which can be condensed back into PET again. A similar technology is underway to depolymerize polyurethane.

IV. *Solvent-Based Processes*: Unlike depolymerization, these processes do not break down polymers with chemical reactions. Suitable solvents are used to dissolve them to remove impurities or contaminants, filter them out, and then reconstitute the polymer. This method seems to work well for recycling polypropylene (PP).

V. *Gasification:* This is a process suitable not only for some plastics but also for municipal waste such as paper and textiles. The waste mixture is heated and gasified under low-oxygen conditions to form a mixture of hydrogen and carbon monoxide and eventually can be made to yield ethanol.

In addition, though PE is recyclable, the energy-intensive process yields only low-value products or "down polymers." Scientists at U.C. Santa Barbara, University of Illinois, and Cornell University have recently developed a catalytic process to up-cycle HDPE into something of higher value—alkyl aromatic hydrocarbons (Zhang et al. 2020), which can be used as a feedstock for other, novel polymers.

## 10.8   THE PROBLEM OF MICROPLASTICS

In 1972, scientists from Woods Hole Oceanographic Institute were trawling the surface of the Sargasso Sea in the North Atlantic to collect brown algae seaweed. But their attention was diverted by an unexpected catch of tiny plastic particles. Repeated net tows caught more and more plastic particles. The researchers estimated a patch of about 3500 plastic particles per square kilometer. But no one could forecast that this was a sign of things to come and not an isolated anomaly.

In 1997, while crossing the mid-north Pacific Ocean from California to Hawaii, Captain Charles Moore accidentally discovered a roughly 200-mile diameter soup of plastic litter floating on the surface appearing around 2 inches thick. (Current estimates put the area more nearly at twice the size of the state of Texas and much deeper.) The soup contained a collection of discarded items of all shapes, sizes, and colors: plastic spray nozzles and product bottles, rubber shoe parts, tooth brushes, butane lighters, Nestlé coffee can lids, Styrofoam shards, and fishing nets that defy forensic analysis. No doubt, this had accumulated over several years from passing merchant ships, ferries, passenger ships, fishing vessels, pleasure craft, aquaculture farms, and off-shore oil and gas platforms. It turns out that this was only one of five gyres—slow, rotating whirlpools—accumulating plastics in the oceans. Plastic suspended in salt water chemically degrades significantly more slowly than its counterpart on land. The salt water provides a cooler environment and protects the polymers from both air oxidation and solar radiation. Ocean algae and barnacles coat the surface of the plastic pieces, further protecting it from decomposition. Not only are the plastic shards choking sea animals, but also PCB released from some plastics is poisoning sea life and leading to massive fish kills. It is important to keep in mind that over half of the world's oxygen supply is generated by marine phytoplankton.

Resembling a patchwork quilt, this was probably the origin and initial recognition of the microplastics pollution problem. Most microplastics come from plastic pollution that gets broken down in the ocean and then ingested by marine animals and organisms. Particles are usually identified by manually separating them from animal tissue, suspending the particles in a solvent, and then analyzing them using spectroscopic techniques such as infrared or Raman spectroscopy. Both particle count and particle mass are important to measure and are determined by different analytical techniques. Microplastics are defined as plastic fragments whose size is between 1 micron and 5 mm. Nanoplastics are even smaller particles, less than

100 nm in size. A human hair has an average diameter in the range of 25–50 microns (1 micron = 1 μm = 1000 nm).

Here are some relevant facts:

- The United Nations Environment Program, among many others, calculates that 8 million metric tons of plastic now end up in the ocean or aquatic environment each year. It is estimated that 5.25 trillion plastic particles, many of which are microplastic particles, are currently floating in the world's oceans.
- According to an articles published in Science News, February 20, 2016 and July 6 and 20, 2019, ocean area and depth profiles indicate microplastics have found as deep as 1000 m. Particles less than 5 mm make up most of the plastic debris with concentrations as high as 100,000 particles per sq. kilometer at the surface and average concentrations of 12 particles per cubic meter at depths between 200 and 600 m.
- The U.N.E.P. also estimates that over 63 billion plastic bottles are dumped into oceans and landfills each year. It takes about 450 years for a single plastic water bottle to decompose.
- Plastic debris causes the deaths (by suffocation and ingestion leading to gut obstruction) of more than 1 million seabirds and 100,000 sea mammals each year, including whales, dolphins, and sea turtles.
- Henderson Island, a tiny landmass in the eastern South Pacific, has been found by marine scientists (at the University of Tasmania and the UK's Royal Society for the Protection of Birds) to have the highest density of anthropogenic debris recorded anywhere in the world: 99.8 is plastic, consisting of an estimated 38 million pieces, weighing 17.6 metric tons, 68% not even visible, 4500 items per square meter, with 13,000 new items washing up daily.
- Microbes digest microplastics and inhibit zooplankton growth ("Plastics, Plastics Everywhere," September–October 2019). Video images are available online, showing zooplankton consuming microplastics, subsequently traveling up the food chain to fish ultimately eaten by humans. [See video of zooplankton ingesting microplastic: https://www.youtube.com/watch?v=3vKLokx6iGg–Brief]
- Bacteria such as *Escherichia coli* have been discovered on the surface of microplastics.
- Synthetic hydrophobic contaminants in the ocean, such as bis-phenol A or other toxic monomers, can adsorb to the surface of microplastics.
- Microplastics have been found in the air, freshwater, soil, sea salt, and dust in homes.
- According to research conducted by the Norwegian Institute for Air Research, one source of airborne microplastics is traffic, largely through friction between tires and roadways as well as the wear of brake pads, carried by atmospheric transport throughout the globe.

Researchers at the University of Queensland and University of Exeter used pyrolysis–gas chromatography/mass spectroscopy (GC/MS) to identify and quantify microplastics in treated sewage sludge (Acter et al. 2020). The research team later adapted this method to analyze seafood and identified five common polymers present in marine plastic waste: polyvinyl chloride (PVC), PE, PS, PP, and poly (methyl methacrylate) (PMM). Sardines had the highest concentration of microplastics—a 100 g sample/serving had 30 mg of microplastics—while squid had the lowest concentration—0.7 mg per 100 g sample/serving. PVC was found in all the samples, and PP was found in all but oysters.

According to another study conducted by Austria's Environment Agency and the Medical University of Vienna, up to nine different microplastics sized between 50 and 500 microns (1 micron = 1 µm = $1.0 \times 10^{-6}$ m) were found in human stool, with PP and PET being the most common, at an average concentration of 20 particles per 10 g of stool.

Finally, the mere act of opening plastic packaging, tearing by using hands, scissors, or knives, has been found by Cheng Fang at the University of Newcastle in Australia (Sobhani et al. 2020) to be another source of microplastics and introduction into the environment. Plastic packaging consists of shrink wraps, shopping bags, plastic bottles, gloves, packaging film, or foam. Often, the contents themselves may also consist of additives such as microbeads in soaps, bodywashes, or cosmetic products that are released into the environment and enter the wastewater stream but escape treatment.

# 11 Radioactivity and Nuclear Chemistry

Most atoms that are encountered on a day-to-day basis are stable and are likely to persist for millions of years barring outside intervention. Some atoms, however, are not stable. Recall that the nuclei of atoms are made up of protons and neutrons. Protons are all positively charged and repel each other. For this reason, there is a lot of energy bound up in the nucleus. In some atoms, this energy is too much. These atoms tend to decay over time into other atoms until they turn into an atom that is stable. These processes, however, tend to release particles with significant amounts of energy. Radioactivity, which originates in the nucleus of an atom, is the means by which excess energy is released from unstable, high-energy atoms as they decay into lower-energy atoms that are stable.

## 11.1 TYPES OF RADIOACTIVITY

The stability of atoms depends on several factors, a detailed explanation of which is beyond the scope of this text. The easily observed cause of stability is due to the ratio of neutrons to protons, which is ideally slightly over one, favoring neutrons. Atoms with 84 protons or more are always radioactive. (Bismuth, element 83, is technically radioactive, but its half-life is so long, that for practical purposes it can be treated as stable.) Decays occur in order to move atoms, sometimes through an intermediate, closer to having this more ideal ratio.

To track changes in atomic nuclei, it is important to use notation indicating all parts of an atom. When discussing radioactivity, different isotopes of an atom can behave differently. As an example, Equation 11.1 shows the full notation for uranium-238:

$$^{238}_{92}U \tag{11.1}$$

The preceding superscript indicates the atomic mass or the total number of neutrons and protons in this particular isotope of uranium, 238. The preceding subscript indicates the number of protons, 92. By difference,

$$238 - 92 = 146 \tag{11.2}$$

is the total number of neutrons. Knowing the number of protons and neutrons provides the information needed to recognize and work with atoms during nuclear decay.

Historically, three types of radioactivity were discovered and named in the early 20th century and are now commonly known, though other types are also now understood. These three types of radiation are alpha, beta, and gamma radiation. Alpha and beta radiation are a direct way for atoms to change their atomic mass or atomic number. Gamma radiation is usually a type of radiation that accompanies other decay events.

Alpha particles, with the Greek letter $\alpha$ as the symbol, are composed of two protons and two neutrons. This is the nucleus of a helium atom. The letter $\alpha$ and $_2^4\text{He}$ are interchangeable symbols for an alpha particle. Alpha particles are relatively low-energy particles, and their large size makes it likely that they will collide with something in their way. Alpha particles have a short range in air.

Beta particles, with the Greek letter $\beta$ as the symbol, are high-velocity electrons ejected from the nucleus. They are, interestingly enough, the product of the decomposition of a neutron into an electron and a proton. Beta particles have a longer range in air and can penetrate thin barriers.

Gamma rays are high-energy photons released, usually, as a by-product of a nuclear decay event. Gamma rays have the ability to pass through very large distances in the atmosphere and require lead shielding to be reliably stopped. Gamma rays are also very damaging to organic material. Figure 11.1 depicts these characteristics of radioactive decay particles.

As shown above, alpha ($\alpha$) particles are stopped by paper. Beta ($\beta$) particles penetrate paper, but are blocked by thin metal sheets. Gamma ($\gamma$) radiation can penetrate paper and aluminum, but is attenuated by lead. Thick lead shielding is required to stop all gamma radiation. Image by Wikimedia user Stannered, released under the Creative Commons Attribution-Share Alike 3.0 License. Original can be found at: https://commons.wikimedia.org/wiki/File:Alfa_beta_gamma_radiation_penetration.svg.

Other types of decay include positron decay, as well as electron capture. Positron decay releases a short-lived positron particle, which is a particle like an electron, but with a positive charge. Electron capture is the capture of an electron that is in an orbital in the atom by a proton. The proton and the electron combine to form a neutron. Both of these types of decays are accompanied by the release of a high-energy photon (X-ray or gamma ray).

Nuclear decays generally occur in order to move a nucleus closer to the ideal ratio of protons and neutrons and below atomic number 84. As an example, uranium-238 is well above the atomic number of 84, which is the threshold for stable nuclei.

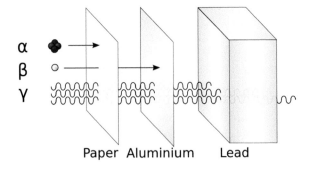

Paper  Aluminium    Lead

**FIGURE 11.1**   Radioactive particle penetration.

Uranium-238 decays by alpha emission, reducing its atomic number, as shown in Equation 11.3:

$$^{238}_{92}U \rightarrow {}^{234}_{90}Th + {}^{4}_{2}He \qquad (11.3)$$

Note that the total of the atomic mass of the products $(234+4=238)$ and the total of the atomic number of the products $(90+2=92)$ are the same as the atomic mass and atomic number of the starting nucleus. That is, mass and the number of particles are conserved in this process. The decay product, thorium, is still not a stable atom. For the full decay pathway, see Section 11.4 in this chapter.

Beta decays tend to occur in order to reduce an abundance of neutrons relative to protons. Although a slight excess compared to 1:1 is preferred for neutrons, there is such a thing as too much. As an example, carbon-14, $^{14}_{6}C$, has eight neutrons and six protons. It undergoes decay as follows:

$$^{14}_{6}C \rightarrow {}^{14}_{7}N + {}^{0}_{-1}\beta \qquad (11.4)$$

Here, the total mass has not changed. Although strange, the effective number of protons also has not changed, as a proton has a (+) charge, whereas an electron has a (−) charge. The negative charge of the electron released can be subtracted from the seven protons in the new nitrogen atom to show conservation of mass and charge. The product atom now has a 1:1 neutron to proton ratio, which is more favorable than the previous configuration.

To be noted here is the significant difference between the commonly encountered types of carbon everyone is familiar with, $^{12}C$ and $^{13}C$. Carbon materials such as graphite or charcoal are made up almost entirely of carbon 12 and carbon 13, neither of which is radioactive. The ratio of protons to neutrons in these two atoms is stable. Carbon 14, however, is radioactive, as the ratio of neutrons to protons is too high. Sometimes atoms of the same element do behave differently in some ways.

## 11.2 HALF-LIFE

Nuclear decay occurs in accordance with first-order kinetics (see Chapter 6, Section 6.3). Each type of radioactive nucleus can be characterized with a specific half-life. One half-life is the amount of time it takes for half of the present material to decay into other nuclei. The half-life is determined by a number of factors, including the excess energy bound up in the nucleus, the stability of the product nuclei, and the ease with which a decay event can occur. Half-lives can be short, on the scale of seconds or less, or exceptionally long, several billion years (and sometimes longer).

The mathematical relationship that can be used to express half-life is

$$\frac{N_c}{N_i} = \left(\frac{1}{2}\right)^{\frac{t}{t_{1/2}}} \qquad (11.5)$$

In this equation $N_C$ is the current amount of material, and $N_i$ is the initial amount of material. On the other side, $t$ is the time that has passed since $N_i$ was observed, and

$t_{1/2}$ is the characteristic half-life of the material being observed. Sometimes, the half-life is unknown, but the rate of the decay is available.

**Example 11.1:** Decay Rate and Dating of Uranium Rock

Using this relationship, we can find the period of time during which a sample has been decaying. One application for this equation is the calculation of the age of a sample. For example, the age of a rock can be dated by the relative amounts of $^{238}U$ and $^{206}Pb$. When rocks are formed, some amount of $^{238}U$ may be trapped within the rock. All of this uranium will eventually decay to $^{206}Pb$, and uranium decay is the only source of $^{206}Pb$. Let us assume a given sample has 230 mg of $^{238}U$ and 120 mg of $^{206}Pb$. The half-life of $^{238}U$ is $4.5 \times 10^9$ years. Mole calculations show that the rock has 0.966 mol of uranium and 0.583 mol lead. Since all of the lead came from uranium, the initial amount of uranium was 1.549 mol. Plugging into the above equation, we get Equation 11.6:

$$\frac{0.966\,\text{mol}}{1.549\,\text{mol}} = \left(\frac{1}{2}\right)^{\frac{t}{4.5\times10^9\,y}} \tag{11.6}$$

$$0.624 = \left(\frac{1}{2}\right)^{\frac{t}{4.5\times10^9\,y}} \tag{11.7}$$

Taking the log of each side of Equation 11.7 (which is frequently done in exponential equations) yields a more manageable equation and easier solution:

$$\log 0.624 = \log\left(\frac{1}{2}\right)^{\frac{t}{4.5\times10^9\,y}} \tag{11.8}$$

$$\log 0.624 = \frac{t}{4.5\times10^9\,y}\log 0.5 \tag{11.9}$$

$$-0.205 = \frac{t}{4.5\times10^9\,y}(-0.301) \tag{11.10}$$

$$t = 3.0\times10^9\,y \tag{11.11}$$

This makes the rock in question $3.0 \times 10^9$ years old. That is a very old rock. The value, however, is reasonable, if we consider the age of the Earth is a billion years more than the age of this very old rock.

## 11.3   UNITS OF RADIATION

There are a multitude of ways to measure radiation. Since radiation can be potentially very dangerous, understanding these units is critical. Although there are two specific SI units to consider, the becquerel and the gray, there are a number of other units that have been used over time to measure either rate of decay or the degree to which something is dangerous. Older documentation often uses these, so they are also discussed.

### 11.3.1 DECAY RATE

The SI unit of becquerels, Bq, measures the rate at which decay events occur within a sample. The becquerel is equal to 1/s or $s^{-1}$, with 1 Bq equal to 1 decay event per second. These decay events are usually detected as alpha, beta, or gamma ray emissions per second. Becquerels are used to distinguish the measurement from other quantities that also have $s^{-1}$ as the unit (such as frequency). Becquerels correspond directly to the amount of compound present in a sample. For a given material, there is a particular rate at which the material decays. There is a linear relationship between the amount of radioactive material and the number of decay events per unit of time.

An older unit measuring the decays per unit of time is the curie (Ci). A curie is equal to the number of decay events observed in 1 gram of radium-226 per second. To convert between the older units, the relationship to use is 1 Ci = 37 GBq.

### 11.3.2 ABSORBED RADIATION

The most important aspect of radiation to be aware of is the rate at which it is absorbed by a human body and the relative effect of different types of radiation. Although decay rate gives an idea of how rapidly particles are generated, not all particles are absorbed by nearby objects, and their health effects might be drastically different. Measurements dealing with these phenomena break into two categories: one dealing with the simple matter of energy absorbed and the other category of units includes in the calculation a factor accounting for the type of particles effect on health.

The SI unit of absorbed radiation is the gray (Gy). One gray is equal to 1 joule of energy absorbed per kilogram of material (1 Gy = 1 J/kg). This can be compared to the sievert (Sv), which, although it shares the same units, means the equivalent energy absorbed. (1 Sv = 1 J/kg)

The relationship between grays and sieverts is governed by definitions put forth by the International Commission on Radiological Protection (ICRP). In Equation 11.12, the term $D$ is the absorbed dose measured in grays, and $H$ is the equivalent dose absorbed measured in sieverts.

$$H = W_R \times D \qquad (11.12)$$

$W_R$ is a factor, determined by the ICRP that accounts for the varying impact of particles on human health. These characteristics are shown in Table 11.1.

**TABLE 11.1**
**$W_R$ Factors for Radioactive Particles**

| Particle Type | $W_R$ |
|---|---|
| EM radiation (X-ray, gamma), beta radiation | 1 |
| Protons | 2~5 depending on energy |
| Neutrons | 2.5~20 depending on energy |
| Alpha particles | 20 |

The above factors show that certain types of radiation cause more damage per unit of energy absorbed. This is due to the way different types of particles cause their damage. Small particles like photons and electrons tend to interact less with tissue, and their energy is dissipated over a greater distance in the body. Heavier particles dissipate their energy much more quickly, colliding with other particles much more frequently in the body, concentrating their energy into a smaller space. This tends to magnify the biological impact of radiation.

**Example 11.2:** Calculating Time of Decay for Contaminated Samples

A major concern with radioactive waste is the length of time for which it has to be stored such that it will have been sufficiently decayed that it is safe. Iodine-131 is used in medicine and is a radioactive waste product of some medical procedures. Appropriate housing for waste of this sort needs to be provided. The half-life of iodine-131 is 8 days. Two liters of medical waste contaminated with $^{131}I$ has to be disposed. Regulations require that the disposed material must have an activity level of 0.008 Ci/m³ or less. The sample currently generates 800 MBq of radiation. How long must the sample be stored in order for it to be safe to dispose of?

*Solution*

As could be expected, the units that regulations are stated in and the units measurements are made in often do not match. First, restate all measurements in common units. Two liters of waste is 0.002 m³ of waste. 800,000 Bq is $2.16 \times 10^{-5}$ Ci. To find the activity per volume, divide the activity by the volume.

$$2.16 \times 10^{-5}\,Ci \div 0.002\,m^3 = 0.0108\,Ci/m^3 \qquad (11.13)$$

The rate of decay, whether in curies or becquerels, is directly proportional to the amount of radioactive material present. For this reason, the half-life equation, Equation 11.5, can be used to calculate the change in the rate of decay. $N_i$ is the initial decay value just calculated, and $N_c$ is the final desired value for disposal to be allowed.

$$\frac{0.008 \; Ci/m^3}{0.0108 \; Ci/m^3} = \left(\frac{1}{2}\right)^{\frac{t}{8d}} \qquad (11.14)$$

Taking the log of both sides yields:

$$\log 0.741 = \frac{t}{8d}\log 0.5 \qquad (11.15)$$

Solving for $t$, the number of days the sample needs to be held before disposal:

$$t = 3.46d \qquad (11.16)$$

Thus this sample would have to be retained for 3.46 days before it would be safe to dispose of.

### 11.3.3 RADIATION EXPOSURES

Radiation is an often misunderstood hazard. It is important to have an appreciation for the relative danger of different exposures. The yearly exposure, for instance, due to radiation is lower near nuclear power plants than coal-fired ones. Coal-fired power plants release a lot of trace radioactive compounds in coal into the atmosphere, whereas nuclear power plants have very tight controls on their emissions. Both of these exposures pale in comparison to medical scans such as dental X-rays or CT scans. In the wake of the Fukushima nuclear accident in 2011, Randall Munroe, the author of the webcomic XKCD, compiled the information presented in Figure 11.2,

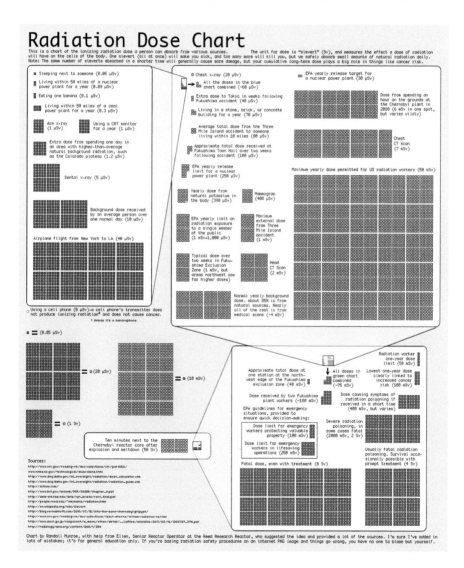

**FIGURE 11.2** Radiation dosage chart.

to give the reader a readily accessible way to understand the relative sizes of various sources of radiation exposure. This chart shows the relative sizes of radiation doses from some common sources of exposure. [It was compiled and generously released into the public domain by Randall Munroe.]

Health effects are best discussed in terms of sieverts, as they measure the impact of radiation on biological systems. In terms of relative size of exposure, 1 Sv is a large dose of radiation to receive, accompanied by a 5.5% increase in risk of cancer as detailed by ICRP. The period of time over which the radiation dose was received has a significant impact on biological effect. An acute 2 Sv exposure can lead to death. As little as a cumulative 0.1 Sv dose over a year has been shown to increase the chance of getting cancer. Close monitoring of any exposure is thus required any time the possibility of exposure exists.

A brief clarification of radiation units may be in order at this point. There are a number of older units used to describe the same measurements as grays and sieverts. Of particular interest are **rads** and **rems**, which appear in older safety guidelines. A **rad** measures delivered radiation and is equal to .01 grays. Rads are not specific to the receiving body. Rems on the other hand are. A **rem** is defined as .01 sieverts. Much like a sievert, a rem is still a fairly large unit of radiation (even though it is smaller than a Sv) and radiation was often measured in mrem or millirem ($10^{-3}$ rem). *As older units, rads have been displaced by grays, and rems have been replaced by sieverts.*

## 11.4    RADIATION IN THE FIELD: DECAY SERIES OF URANIUM

As mentioned at the start of this chapter, nuclear decay occurs because some nuclei are unstable, and decay leads them to a more stable nucleus. This is sometimes a multistep process, with a number of varyingly unstable nuclei being formed until a nonradioactive element is formed. An important example of such a process is the uranium series, the sequence of transformations by which uranium decays into lead. As shown in Figure 11.3, this series has an intermediate product, radon, which is of direct concern in the environment, as radon can become airborne.

While there are trace amounts of other decay events, the majority component of the decay chain is shown in Figure 11.3, and the final product is always lead-206, which is a stable nucleus. There are many alpha decay events, which carry off a great deal of the mass of the uranium nucleus. These decays are the major source of naturally occurring helium gas, as alpha particles eventually pick up electrons and form helium atoms. Beta decays also occur, and these decays adjust the proton to neutron ratio within the atoms formed, moving them closer to a more stable ratio.

Of particular interest is radon-222. Due to the presence of uranium in certain rock formations, a certain amount of radon will be constantly generated by the above decay scheme. Radon is a gaseous element and, if it is close enough to the surface when formed, can rise into the air. The half-life of radon-222 is 3.8 days, which means that it has sufficient time to diffuse out of rock formations, but also decays relatively quickly, making it a potential danger if allowed to concentrate.

Radon is a normal part of the background radiation present all around people on a daily basis. Under some circumstances, however, the gaseous radon can accumulate

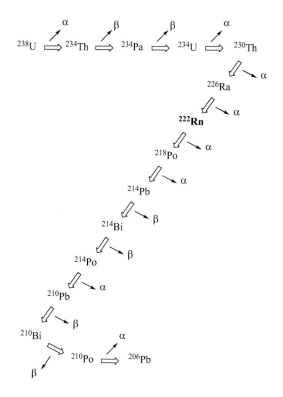

**FIGURE 11.3**   Uranium-238 decay chain. (Note: radon-222, bolded, is a product along this decay chain.)

in enclosed spaces, building up to potentially dangerous concentrations. As radon is gaseous, and radon decays by alpha emission, it has a very high potential to cause radiation poisoning. Since it can be inhaled, internal exposure to radiation is particularly likely. Moreover, it decays into a series of other atoms, which, not being gaseous, can deposit in the lungs, and have short half-lives on the order of minutes. This leads to a great deal of potential radiation exposure from radon and its decomposition products.

The World Health Organization (WHO), the US Environmental Protection Agency (EPA), and the European Union (EU) have slightly different recommendations for the permissible concentration of radon in buildings. These limits are usually defined in terms of decays per second in a particular volume. Depending on the agency, this might be expressed as becquerels per cubic meter or curies (often picocuries or $10^{-12}$ curies) per liter. The conversion here is 1 pCi/L = 37 Bq/m$^3$. The value of 37 Bq/m$^3$ means that every second 37 radioactive nuclei decompose in every cubic meter of air present.

According to WHO, based on the aggregate of a number of long-term studies, there is a roughly 10% increase in the rate of lung cancers for every 100 Bq/m$^3$ present in a home. The EPA suggests action at 2 pCi/L and strongly recommends action at 4 pCi/L. These values correspond approximately to 100 Bq/m$^3$. EU recommendations for action to improve home conditions are at slightly higher 200–400 Bq/m$^3$. The reason that ranges for permissible amounts of radon are often given is due to the potential cost/benefit of any repair to be done to a home. The primary way to alleviate a radon problem is improved ventilation of basement areas, although the EPA and other agencies have publications on a number of ways reduction in radon concentrations in the home can be accomplished.

# 12 The Atmosphere and the Chemistry of Air

Air is an important environmental resource. The chemistry of air and air pollution can be a complex and involved topic. Air is predominantly made up of gases, and reviewing Section 1.9 may help in better understanding some of these topics. Here the details of the atmosphere are broken down and manners in which human activity can alter air quality are discussed, and the impact these effects can have on the environment is reviewed.

## 12.1 THE COMPOSITION OF THE ATMOSPHERE

The atmosphere on the Earth is a complex layer of gases held on to by the gravitational pull of the planet. The atmosphere is generally described in several layers, as the behavior and composition of the gases change the further from the ground they are. The troposphere is the lowest region of the atmosphere, the layer in which people live, up to about 12 km. Above that is the stratosphere, up to about 50 km. Jet airplanes generally fly in this region of the atmosphere. Temperatures in the troposphere decline to around −55°C (−66°F). Contrary to intuition, however, the temperature in the stratosphere increases as the elevation increases, to near 0°C (32°F), due to increased absorption of UV radiation in this region, in particular by the ozone layer. Above the stratosphere is the mesosphere, reaching to about 100 km in altitude. Gases in this region are particularly sparse and temperatures plummet very low. Above these regions are the thermosphere and the exosphere, but these are of little relevance to the discussion here. A diagram of these properties can be seen in Figure 12.1.

The atmosphere is composed primarily of three gases, nitrogen, $N_2$, oxygen, $O_2$, and argon, Ar. Nitrogen makes up ~78% of the atmosphere, $O_2$ ~21%, and Ar ~1%. Roughly 0.04% of the atmosphere is made up of trace gases, mostly $CO_2$, but also methane, hydrogen, and other noble gases.

Air will also contain some small fraction of water in the form of humidity. The maximum amount of humidity is dependent on temperature and is referred to as saturation humidity. Humidity is generally measured in terms of relative humidity. Relative humidity is the ratio of the density of water vapor in air to the saturation density at the current temperature. Table 12.1 lists vapor densities of water at selected temperatures.

$$relative\,humidity = \frac{measured\,density}{saturation\,density} \qquad (12.1)$$

If humidity rises above this point, precipitation as dew or rain occurs.

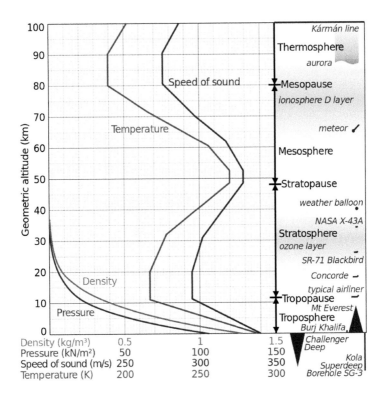

**FIGURE 12.1** Properties of the atmosphere at changing altitudes. (Image courtesy of Cmglee released under Creative Commons Attribution Share-Alike 3.0 license. The original image may be found here:https://upload.wikimedia.org/wikipedia/commons/archive/9/9d/20141031174254%21Comparison_US_standard_atmosphere_1962.svg.)

## TABLE 12.1
### Density of Water Vapor in Air

| Temperature °C | Vapor Density g/cm³ | Temperature °C | Vapor Density g/cm³ | Temperature °C | Vapor Density g/cm³ |
|---|---|---|---|---|---|
| 10 | 9.4 | 22 | 29.4 | 34 | 37.6 |
| 11 | 10.0 | 23 | 20.6 | 35 | 29.6 |
| 12 | 10.7 | 24 | 21.8 | 36 | 41.8 |
| 13 | 11.4 | 25 | 23.0 | 37 | 44.0 |
| 14 | 12.1 | 26 | 24.4 | 38 | 46.3 |
| 15 | 12.8 | 27 | 25.8 | 39 | 48.8 |
| 16 | 13.7 | 28 | 27.2 | 40 | 51.2 |
| 17 | 14.5 | 29 | 28.7 | 41 | 53.8 |
| 18 | 15.4 | 30 | 30.4 | 42 | 56.6 |
| 18 | 16.3 | 31 | 32.0 | 43 | 59.5 |
| 20 | 17.3 | 32 | 33.8 | 44 | 62.5 |
| 21 | 18.3 | 33 | 35.7 | 45 | 65.5 |

**Example 12.1:** Calculating Air Density Based on Humidity

The relative humidity is reported to be 80% on a hot day. It is 30°C outside. What is the current density of water in air? What percent by mass of the air is water?

*Solution*

At 30°C (86°F), the density of water at saturation is 30 g/m³. If relative humidity is 80%, then:

$$NO_2 + photona\ NO + O^{0.80} = \frac{x\,g/m^3}{30\,g/m^3} \text{ therefore } x = 24\,g/m^3 \tag{12.2}$$

There is 24 g/m³ of water in the air. The density of dry air at 30°C is 1164 g/m³, which is assumed to be a close enough value for the second calculation.

$$\frac{24}{1164} = 0206 = 2.06\% \tag{12.3}$$

The air in this case is comprised of roughly 2% water by mass.

## 12.2 DEFINITIONS OF GASES, FUMES, VAPORS, AEROSOLS, AND MISTS

Although in everyday use the terms gas, fume, vapor, aerosol, and mist are often used interchangeably with one another, it is important to note that these terms have specific, scientific, or technical definitions.

The term gas refers to one of the three common states or phases of matter. In gases, individual particles of matter, be they molecules, atoms, or ions, are spaced far apart from each other, such that they experience little to no intermolecular forces. These particles are largely invisible and move randomly within the gas phase and interact very little with other particles in the gas.

The technical definition of a vapor is a gas that is below its critical temperature. The critical temperature is the temperature above which a material cannot be compressed into a liquid. In more common use, vapor is often used to refer to a liquid that has formed in suspension in air from having been recently evaporated, such as the vapor above boiling water.

An aerosol is a gas–liquid or gas–solid colloid. Colloids (discussed in Chapter 3) are materials that form a stable system, but the particle sizes of the solute are large in comparison to the solvent molecules. Aerosols are very finely dispersed liquid or solid particles that hang in suspension in air. They are usually 100 microns or less in diameter, e.g., as produced in a human sneeze or a cough. Those that are larger than 300 microns are usually droplets. Ash released by power plants, referred to as fly ash, is one example of a gas–solid colloid. The paint sprayed from a spray paint gun can also form an aerosol while in the air. There is a fine distinction between colloids and suspensions. A colloid is a stable system that does not require periodic mixing. Milk and blood are examples of colloids, also referred to colloidal dispersions. A colloidal suspension, on the other hand, requires periodic stirring to approach uniformity. Paints and many cough medicines, which require stirring or shaking, are examples of suspensions.

The terms dust, fume, and mist have technical definitions when talking about air quality. Mist is used to describe water–air aerosols generated either chemically or mechanically. Fumes are fine, particulate solid–air aerosols generated through burning or condensation. Tobacco smoke is composed of fumes. Dust is composed of small solid particulates, visible for the most part, that are forced into air suspension through mechanical means such as wind.

## 12.3 PRIMARY AIR POLLUTANTS FROM STATIONARY AND MOBILE SOURCES

Human activity generates many sources of contaminants in the atmosphere, what is generally termed "pollution." The main source of most of these is combustion of fossil fuels for energy, though other industrial activities also generate large quantities of certain pollutants.

### 12.3.1 CARBON DIOXIDE

Carbon dioxide, $CO_2$, is of significant environmental concern for a large number of reasons. A discussion of atmospheric considerations is found in Chapter 13, where $CO_2$ impact on oceans is discussed. Carbon dioxide is not only emitted as a natural process of life, but is also emitted in significant amounts by industrial activities conducted by humans. Transportation, energy generation, and even other industrial processes such as smelting of iron or calcination in the production of Portland cement are all significant sources of anthropogenic carbon dioxide.

Carbon dioxide emissions are generally measured in giga metric tons per year (Gt/y). It is important to pay careful attention to the definition of the measurement, as sometimes it is reported as carbon dioxide and sometimes as carbon. The important distinction is that sometimes the entire mass of the molecule is considered (as carbon dioxide), and sometimes only the carbon part is considered (as carbon). The conversion between the two measurements is the ratio of the molar masses of $CO_2$ and C, which is 44:12.

### 12.3.2 SOURCES OF CARBON DIOXIDE

According to the United States Department of Energy, 33 Gt of $CO_2$ was released by human activity in 2010 worldwide. There are several major sources of carbon dioxide emissions. The first and foremost is the combustion of fossil fuels. The vast majority of energy needs, for both transportation and electricity, are met by the combustion of coal, oil, or natural gas. In addition, a significant amount of carbon dioxide is also released by other industrial activity. The process by which cement is manufactured and the processes for the purification of many metal ores all generate $CO_2$ as a by-product, in addition to the energy needs of these activities.

The combustion of fossil fuels generates a highly desirable product (heat) and a fairly undesirable by-product ($CO_2$) and a mostly harmless by-product ($H_2O$). A typical reaction for octane (the main component of gasoline) is shown by Equations 12.4 and 12.5:

$$C_8H_{18} + O_2 \rightarrow 8\,CO_2 + 9H_2O \tag{12.4}$$

$$\Delta H_{rxn} = -5460\,\text{kJ/mol octane} \tag{12.5}$$

Some further calculation can show that the amounts of energy carried per unit volume and per unit mass of fossil fuels are both quite high compared to other ways to transport energy, except for nuclear fuels. Their high energy content and the ease with which the energy can be recovered through combustion make fossil fuels ideal for many applications including energy generation and transportation.

A key step in cement production is calcination, wherein calcium carbonate is heated, releasing $CO_2$ and forming calcium oxide as shown in Equation 12.6. The calcium oxide produced by this process is then mixed with a variety of other oxides and other additives to form cements for various applications.

$$CaCO_3 \rightarrow CaO + CO_2 \tag{12.6}$$

Ore processing is also carbon-intensive. Iron ore can be composed of a variety of iron oxides, but a representative reaction for the smelting of iron ore is shown in Equation 12.7:

$$Fe_2O_3 + 3C \rightarrow 2\,Fe + 3CO_2 \tag{12.7}$$

Iron and other ores can be reduced to their elemental form in high-temperature reactions in the presence of carbon and in the absence of oxygen. The oxygen atoms can be transferred to carbon atoms, and escape as gas, leaving behind molten iron. This process, called smelting, is central to metal production.

As can be seen above, key processes to the modern world, energy production, transportation, and construction, all generate carbon dioxide. It is important to recognize that while critical to the human way of life, these processes have a direct impact on the human environment. While this text does not allow for a complete and full discussion of the impact of carbon dioxide on the environment, the following sections discuss some of the science relating to this topic.

### 12.3.3 CARBON DIOXIDE AND GLOBAL WARMING

The problem with burning fossil fuels is that the carbon dioxide released into the environment has an impact on the global climate. Although the amount of $CO_2$ released in any given year is relatively small compared to the total $CO_2$ in the atmosphere, humans have been releasing dramatically increasing amounts of carbon dioxide since the beginning of the Industrial Revolution in 1750. According to the International Panel on Climate Change (IPCC), in 260 years, humanity has managed to add an impressive amount of carbon dioxide to the atmosphere a little bit at a time. Ice core samples from the Antarctic have established that atmospheric $CO_2$ concentrations have been between 200 and 300 parts per million (ppm) for at least the last 800,000 years. As of May 2015, the concentration of $CO_2$ in the atmosphere has reached 404 ppm. There is a reason for concern.

Carbon dioxide acts as an insulator in the atmosphere. The sun irradiates the Earth with a very large amount of energy. Not all of this energy is absorbed, however. A significant fraction of the energy that reaches the ground is reflected back away from the Earth. Most of this reflected thermal energy is radiated into space. The incoming radiation from the sun, however, has a different profile than the outgoing radiation from the Earth. Most of the energy from the sun is in the ultraviolet (UV) and visible energy range, whereas the Earth radiates in the infrared (IR) range of the spectrum (see Figure 12.1). While $CO_2$ does not interact with incoming light to a significant degree, it interacts with outgoing IR radiation well.

The various components of the atmosphere tend to absorb light as vibrational energy. Chemical bonds in molecules are not rigid, but are rather like springs. They can vibrate in a number of different ways, oscillating in length, wiggling side to side, changing the exact angle of a particular bond from moment to moment. Each of these vibrations corresponds to a particular amount of energy, which that bond can absorb most readily. The preferred absorbances of some molecules in the atmosphere can be seen in Figure 12.2.

A large component of the ability of the atmosphere to retain heat is actually due to water. In fact, this heat retention is an important component of the ability of the planet to support life. Carbon dioxide absorbs in a region that water does not absorb in and thus significantly increases the heat retention of the atmosphere. As carbon dioxide concentrations increase, the amount of heat retained by the atmosphere also increases. The net effect is an increase in global temperatures. Although other gases (particularly methane) also increase this heat retention effect, $CO_2$ is currently the biggest and most worrying component of atmospheric contaminants.

### 12.3.4   SOx AND NOx

Two classes of atmospheric contaminants, SOx and NOx, are also sourced from the combustion of fossil fuels. SOx is a reference to a class of airborne sulfur oxides with a varying number of oxygen atoms, and NOx is a class of nitrogen oxides, again with a varying number of oxygen atoms. Fossil fuels are predominantly made up of carbon and hydrogen, but do also contain small amounts of other elements. In this case, the concern is with sulfur and nitrogen, each of which makes up around 1% of the mass of most fossil fuels, although the exact values will vary depending on the exact quality, source, and type of fossil fuel (coal, petroleum, or natural gas).

SOx are a result of the oxidation of sulfur atoms that are present in small quantities in burned fuels. Although SO and $SO_3$ can both be formed, the main component at the point of release of SOx pollution is $SO_2$. Unlike $CO_2$, $SO_2$ is not a greenhouse gas and in fact tends to reflect solar energy out of the atmosphere, thus acting to counter greenhouse effects. However, chemical processes that occur in the atmosphere convert $SO_2$ into other compounds of concern.

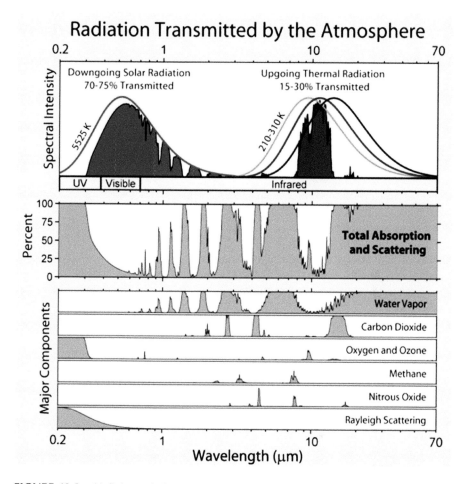

**FIGURE 12.2**   (a) Solar and planetary energy emissions, and components absorbed by the atmosphere. (b) Percent of energy radiation absorbed by atmosphere. (c) Contributions from various atmospheric gases and effects. (Image courtesy of Robert A. Rhode released under Creative Commons Attribution-Share Alike 3.0. The original image may be found here: https://commons.wikimedia.org/wiki/File:Atmospheric_Transmission.png.)

$$SO_2 + H_2O \rightarrow H_2SO_3 \qquad (12.8)$$

$$2SO_2 + O_2 \rightarrow 2SO_3 \qquad (12.9)$$

$$SO_3 + H_2O \rightarrow H_2SO_4 \qquad (12.10)$$

The eventual fate of SOx in the atmosphere is to form sulfurous acid or sulfuric acid. These compounds are a major source of another environmental hazard, acid rain. Acid rain has been responsible for an impact both on the biome and on man-made structures. The impact of acid rain has been noticeable on a number of buildings that were made from marble or limestone, as the acidity of the rain is sufficient to very slowly dissolve these materials. Limestone and marble are both made up of calcium carbonate and can go through an acid–base reaction with acid rain as per Equation 2.11:

$$CaCO_3 + H_2SO_4 \rightarrow CaSO_4 + H_2O + CO_2 \qquad (12.11)$$

Acid rain can also change the pH of bodies of water as well as that of the soil. Many animals are sensitive to such changes and cannot survive these kinds of changes to their environment. Sulfur emissions have been fairly successfully controlled by government regulation.

NOx refers to NO and $NO_2$ as atmospheric impurities. NOx have both natural and man-made sources. Although $N_2$ and $O_2$ do not react with each other under ambient conditions, if sufficient energy is supplied, they will react to form NO and $NO_2$. In the environment, this occurs most often in lightning storms. Any combustion process occurring at a high enough temperature will generate these by-products, so the combustion of fossil fuels in power plants and even in vehicles produces some NOx. The small amounts of N atoms in fossil fuels also tend to be converted to NOx during combustion.

Much like SOx, the eventual fate of NOx is to become nitric acid through reactions with atmospheric oxygen and water.

$$4\,NO + 3\,O_2 + 2\,H_2O \rightarrow 4\,HNO_3 \qquad (12.12)$$

$NO_2$ goes through several steps, eventually forming nitric acid as well. The overall process for $NO_2$ oxidation to $HNO_3$ can be written as follows:

$$4\,NO_2 + O_2 + 2\,H_2O \rightarrow 4\,HNO_3 \qquad (12.13)$$

Much like SOx, NOx also contributes to acid rain, as its conversion to nitric acid would imply. NOx also have a part in photochemical smog and can be a part of the chemical cycle involving ozone, both of which are discussed in the following sections.

Dealing with SOx and NOx pollution is best achieved by control at the source of emission. SOx emissions are best controlled by use of fuels that are low in sulfur content. Sulfur can be removed from flue gas chemically either with treatment with NaOH or with CaO if necessary, for example, as seen in Equation 12.14 for calcium oxide. Calcium sulfite is a solid, and thus the sulfur is removed from the emission gases.

$$SO_2 + CaO \rightarrow CaSO_3 \qquad (12.14)$$

NOx are formed under higher-temperature combustion processes. Economics compete with a reduction in temperature, however, as energy generation becomes less efficient at lower temperatures. NOx are often treated chemically using selective catalytic reduction (SCR). In this reaction, ammonia, or some other nitrogen-based reducing agent, is used to affect the conversion of NOx to more benign compounds, an example of which is shown in Equation 2.15:

$$NO + NO_2 + 2NH_3 \rightarrow 2N_2 + 3H_2O \qquad (12.15)$$

SCR has mostly been applied to emissions from large-scale processes such as power plants and waste processing plants, but has also been scaled down and applied to treat emissions from diesel engines used in ships, trucks, and automobiles.

## 12.3.5  OZONE

Ozone is a toxic gas essential for life on the Earth. There is a layer of the atmosphere, referred to as the ozone layer, in which ozone is formed by the absorption of high-energy UV rays and thus serves to protect organisms living on the Earth from their potentially harmful effects. Ozone can sometimes also be formed closer to ground level, in which case it can be a potential hazard, as discussed in the next section.

The ozone layer is found 20–30 km above ground level in the stratosphere. In this region of the atmosphere, high-energy photons (UV and even some X-rays) are absorbed. Photons are the basic subunits of electromagnetic radiation. One photon is the minimum amount of energy that can be delivered by light of a particular wavelength. UV and X-ray are high-energy photons, higher than visible light. The photochemical process is shown in Equations 12.16–12.21.

$$O_2 + high\,energy\,photon \rightarrow 2O \qquad (12.16)$$

$$O + O_2 \rightarrow O_3 \qquad (12.17)$$

Thus, the formation of ozone allows for the capture of some portion of UV radiation incoming from the sun. The ozone in the ozone layer continues to react with incoming UV radiation, absorbing more photons and then rapidly reforming ozone:

$$O_3 + photon \rightarrow O_2 + O_4 \qquad (12.18)$$

$$O + O_2 \rightarrow O_3 \qquad (12.19)$$

Ozone is also lost from this system if an ozone molecule reacts with oxygen atom or if oxygen atoms react with each other.

$$O_3 + O \rightarrow 2O_2 \qquad (12.20)$$

$$O + O \rightarrow O_2 \qquad (12.21)$$

Thus the concentration of ozone remains relatively low but allows for the continued absorption of medium to high-energy UV radiation. This balance can be disrupted by certain atmospheric pollutants. Chlorofluorocarbons (CFCs) are a broad class of organic compounds that are heavily chlorine and fluorine substituted. These compounds have seen broad use in refrigeration and as aerosol propellants until their negative impact on the ozone layer was discovered. CFCs tend to be volatile and long-lived due to their chemical inertness. In the upper atmosphere, however, they can be decomposed to form chlorine radicals.

$$CCl_4 + photon \rightarrow Cl + CCl_3 \tag{12.22}$$

Chlorine radicals are highly reactive and will quickly consume available ozone.

$$Cl + O_3 \rightarrow ClO + O_2 \tag{12.23}$$

$$Cl + O_3 \rightarrow ClO + O_2 \tag{12.24}$$

This cycle regenerates the Cl that the reaction started with and is thus catalytic in nature. The Cl radical generated from the CFC molecule can decompose many ozone molecules. The Montreal Protocol, established in 1987, is an international agreement to limit the production and use of CFCs. Despite this, some production and use of CFCs do continue to this day.

### 12.3.6 PHOTOCHEMICAL SMOG

A photochemical smog is a mixture of pollutants that are generated by the action of solar radiation on emissions from the combustion of fossil fuels. The chemistry that occurs when driven by high-energy photons aggravates the toxicity and corrosiveness of the emissions of power plants and automobiles. In particular, there are three components needed to generate a photochemical smog; those are NOx, volatile organic compounds (VOCs), and photons. VOCs are organic compounds in the atmosphere originating from the partial combustion of fossil fuels, as well as the evaporation of fuels and other organic compounds such as paint thinners.

The root of the problem here is the reaction that $NO_2$ undergoes with photons to generate oxygen radicals.

$$NO_2 + photon \rightarrow NO + O \tag{12.25}$$

The oxygen radicals generated by this process are highly reactive. They can go on to react with oxygen in the atmosphere or VOCs and form various components of the photochemical smog.

The reaction of oxygen radicals can be seen in the previous section on ozone and results in the generation of ground-level ozone concentration that can be potentially unsafe. Ozone is strongly oxidizing and can have deleterious effects on the respiratory system of people and animals, as well as corrosive effects on metals or plastics.

The reactions of oxygen radicals with VOCs can generate a variety of potentially toxic compounds. VOCs are predominantly short-chain hydrocarbons, often with one to three but possibly up to six or even more carbons. Reactions between oxygen radicals and VOCs can generate aldehydes. These aldehydes can go on to react with $NO_2$ to generate peroxyacyl nitrates, shown in Figure 12.3.

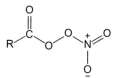

**FIGURE 12.3**   General formula of peroxyacyl nitrates.

Peroxyacyl nitrates are as a group known as PAN. These molecules are somewhat more stable than ozone but still corrosive and potentially toxic. Their stability also poses a hazard, as it allows for the aerial transport of PAN further from the point of generation. This allows oxidative damage from photochemical smog to reach areas at significant distances from urban areas.

### 12.3.7   FLY ASH AND RELATED ENVIRONMENTAL HAZARDS

The combustion of coal poses another hazard, in the form of very fine particulate material, fly ash. There are strict regulations related to and significant technology has been developed to reduce the release of fly ash. The particles that compose fly ash are micrometer-scale particles of noncombustible materials such as silicates and aluminates left over after the combustion of coal. Though they can be electrostatically filtered from the flue gas from the power plant, the filtration is not 100% effective. The health impacts of fly ash are still debated.

A significant concern with coal combustion is that there are trace elements that are present in coal, which often end up released into the environment in the flue gas. About 50 tons of mercury is emitted into the atmosphere on a yearly basis in the United States due to coal combustion. Mercury has a health impact on animal and human populations. It tends to be readily absorbed by microbes. Since animals have little ability to remove mercury from their system, mercury accumulates in the food chain. Fish from mercury-contaminated water can be particularly hazardous to human health.

Another curiosity related to coal is that it releases large amounts of radioactive waste. The concentration of uranium in coal ranges from 1 to 100 ppm depending on the source and quality of the coal. This appears to be a small number, but combined with yearly coal consumption, a scary figure emerges. The global use of coal in 2012 was 7.4 Gt. Even at 1 ppm, that is 7400 metric tons of uranium generated as waste product. Thankfully this is usually not released and is very dilute in the ash waste, but it is enough to be of concern when considering coal burning.

Fly ash does, however, have some industrial uses. Captured ash can be used in the production of cement, as well as in several other construction products such as bricks, concrete, grout, and general filler material.

## 12.4   INDOOR AIR QUALITY (IAQ)

Indoor air pollutants vary considerably in type and molecular dimensions or particle size compared with outdoor, ambient air contaminants. Furthermore, gaseous pollutants can often piggyback onto particulates. A glance at Figure 12.4 surveys the

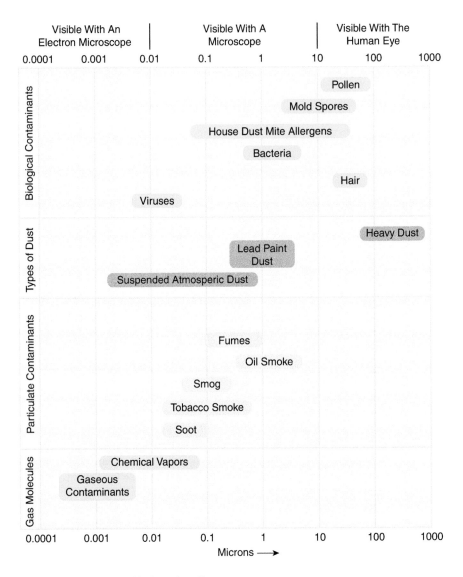

**FIGURE 12.4** Profile of indoor air pollutants.

different types along with their average dimensions. The vertical scale shows various categories or types of contaminants, while the horizontal scale shows size in units of micrometers, i.e., μm or $1 \times 10^{-6}$ m, and visibility ranges with the human eye, an ordinary microscope, and an electron microscope. The horizontal demarcations in size are approximate.

As energy costs rose to heat and air-condition commercial buildings and private homes, structures starting in the late 1970s and early 1980s were designed and constructed with energy efficiency in mind. This meant a tighter assembly with fewer air leaks or indoor–outdoor air exchanges. At the same time, more and more synthetic building and construction materials were also employed—plywood and wafer boards in framing, particle board for indoor furniture, sealing caulks, adhesives, and wall-to-wall carpets. Most, if not all, of these items contain synthetic compounds, which, over time, "off-gas" or "out-gas" into the ambient air. The consequence is a compromised indoor air quality (IAQ), originally referred to as "sick building syndrome." Off-gassing or outgassing simply refers to the release of gases or vapors during the aging and degradation of a material. Estimates based on informal surveys show that of the five million nonresidential buildings in the United States today, 20%–30% are problem buildings, and less than 80% of the occupants are satisfied with their air quality.

Modern materials such as paints, synthetic carpets, curtains, wall paneling, and furniture made of particle board have been implicated. This outgassing is a universal component of indoor air pollution and can easily result in sinus and lung irritation. Formaldehyde is the most studied and ubiquitous of all the VOCs. To make matters worse is a phenomenon known as "the sink effect." An unpainted panel of gypsum, i.e., wallboard, plaster board, or sheetrock can absorb and later re-emit VOCs out-gassed from materials stored in a room, even after those same materials have been removed from the room. Gypsum acts as a sink and is a much stronger sink for form-aldehyde than for any other VOC.

NASA routinely tests air quality of flight cabins aboard its space crafts. During one of its simulated flights, it tested the effects of outgassing of a new adhesive with attractive physical properties used in that cabin. In a 72-hour period with the ambient temperature at 120°F, the adhesive released the following gases and vapors: carbon monoxide, C-5 saturated and unsaturated hydrocarbons, formaldehyde, acetaldehyde, benzene, methanol, ethanol, methyl benzene (or toluene), 2-propanol, 2-butanone, and hexamethyl cyclotrisiloxane. While a temperature of 120°F for a continuous 72-hour period may seem unrealistic, the test does represent what can happen under accelerated conditions of time and temperature. Perhaps, for example, this tempera-ture might be approached for a short period of time in the cabin of an automobile with its windows closed, parked under the sun in the summer, or in the attic of a house, where the event may be repeated several times for the duration of the summer.

## 12.5   EIGHT MAJOR INDOOR AIR POLLUTANTS

There are currently eight major indoor air pollutants: carbon dioxide (already dis-cussed above) is usually not a serious indoor air pollutant, assuming proper venti-lation is followed), carbon monoxide, formaldehyde (the chief VOC, though there are many others as the NASA test result above indicates), nitrogen dioxide, ozone, particulate matter, radon, and sulfur dioxide.

**Carbon monoxide**, like carbon dioxide, sulfur dioxide, and nitrogen dioxide, is a gas that is created by the combustion of fossil fuels. Carbon monoxide, CO, is produced by the incomplete combustion of wood, coal, natural gas, oil, gasoline, kerosene, and tobacco. It is colorless and odorless. It is especially toxic because hemoglobin, the transporter molecule in the blood, preferentially binds with carbon monoxide (to form carboxyhemoglobin) instead of the oxygen molecule (to form oxy-hemoglobin). In fact, carbon monoxide has an affinity for hemoglobin that is roughly 200 times stronger than oxygen. The net result is oxygen deprivation for the human body. The immediate symptoms are impaired brain and vision function, followed by irregular heartbeat, headaches, nausea, weakness, fatigue, and eventual death.

The EPA ambient air quality standards require CO exposure concentrations to be below 9 ppm ($10 \text{ mg/m}^3$) for an 8-hour period and below 35 ppm ($40 \text{ mg/m}^3$) for a 1-hour period. However, carbon monoxide concentrations up to 100 ppm can be tolerated for a few hours without ill-effect.

Noticeable symptoms, however, start to develop when CO concentrations persist at or above 600 ppm for an hour or more. Headaches and fatigue usually develop at about 1000 ppm. While not fatal after 1 hour, this concentration will most likely result in death after 4 hours of exposure. If the concentration continues to increase, symptoms become more severe, and death eventually results at concentrations anywhere from 2000 to 4000 ppm.

Fortunately, over time, carbon monoxide is converted to carbon dioxide by ambient oxygen in the air, so global concentrations of CO do not rise. For comparison, the OSHA permissible exposure limit for indoor exposure to carbon dioxide in a room is 5000 ppm.

**Formaldehyde**, $CH_2O$, as previously mentioned, is ubiquitous, in part because it is so inexpensive to produce and can be used in a wide array of applications. With nearly 10 billion pounds being produced annually, it is difficult to avoid.

It is released from such products as permanent press fabrics, insulation, paints, shampoos, plastics, carpeting, and adhesives used to make plywood and particle board for furniture—urea formaldehyde glue (UF) for indoor furniture and phenol formaldehyde (PF) glue for outdoor furniture. Studies show that the inexpensive UF glue off-gases nearly ten times more formaldehyde than PF glue per unit volume.

Formaldehyde is a colorless, flammable gas at room temperature, with a strong, pungent odor at high concentrations. It is also known as formalin (a mixture of 37%–50% formaldehyde and water with 10% methanol added to inhibit polymerization), methyl aldehyde, or methylene oxide. In the air, it is gradually oxidized to formic acid. Symptoms of exposure vary considerably but include burning of eyes, tightness in chest, headache, asthma attacks, depression, and even death. On average, symptoms start to occur at concentrations from 0.10 to 1.0 ppm. Individuals more sensitive to formaldehyde show symptoms at exposure concentrations as low as 0.03 ppm. Many healthy individuals can tolerate short-term exposures up to 0.20 ppm. Death may occur upon exposure to extremely high concentrations of 50–100 ppm. It has been established as a mutagen and an animal carcinogen, but its relationship to cancer in humans is not completely understood.

**Nitrogen dioxide**, $NO_2$, is another combustion-related gas. Its sources and effects are similar to sulfur dioxide. Nitrogen dioxide has a reddish-brown color, a pungent odor, and is highly toxic. Health effects include burning and choking sensations upon breathing in the upper respiratory tract, as well as irritation of the eyes and skin.

Exposure to nitrogen dioxide at a concentration of 50–100 ppm for a few minutes to an hour can cause lung inflammation, which may last for several weeks. Exposure to concentrations of 150–200 ppm causes a condition known as bronchiolitis fibrosa obliterans, which is fatal in 3–5 weeks. At a concentration of 500 ppm, a condition known as "Silo Fillers Disease" develops and is fatal in 2–10 days.

Fortunately, pollution concentrations of nitrogen dioxide are usually low and in the 1–3 ppm range. At those concentrations, it is of slight concern to humans, although there is some evidence to suggest that chronic exposure at even low concentrations is cumulative and may have long-term effects on the human immune system. It is much less toxic to plants than sulfur dioxide. The EPA ambient air quality standards set a limit of 0.05 ppm (100 ug/m³) for an average annual exposure level.

**Ozone**, $O_3$, is a naturally occurring allotrope (different molecular structure of the same element) of oxygen that exists in the stratosphere. As stated earlier in this chapter, its purpose there is to block out harmful ultraviolet radiation and prevent it from reaching the Earth's surface. It is vital for health. In the troposphere, however, where human and other biological life and activities thrive, it is toxic. It is one of the major components of smog, produced when VOCs and NOx react in the presence of sunlight. VOCs and NOx are emitted from industrial sources, internal combustion engines, and many commercial operations. Ozone can make breathing quite difficult, especially for children and older adults.

Ozone is a highly reactive, bluish gas with a pungent odor. It is a strong oxidizing agent, which can cause coughing, choking, headaches, and fatigue. Other symptoms include eye and mucous membrane irritation. Ozone can be created indoors by sparks from electric motors, copying machines, laser printers, and negative ion generators. Set in 2008, the EPA ambient air quality standard for ground-level ozone was 75 ppb. As of October 2015, EPA tightened this standard to 70 ppb, though many air quality experts advise tightening it further to 60 ppb. OSHA requires exposure concentrations to be below 0.12 ppm (240 ug/m³) for an 8-hour period. Above levels of 0.15 ppm, most individuals start experiencing some sort of discomfort such as coughing, shortness of breath, and nose and throat irritation. Ozone is very irritating to the lungs, which respond by producing a fluid, which fills up the alveoli (tiny air sacs) in the lung. As a powerful oxidizing agent, it can attack lung tissue by oxidizing it.

Besides health effects on humans and other animals, ozone is a more severe problem for plants. Ozone attacks plants and produces identifiable yellow spots on their green leaves. Plants are so sensitive that even a brief exposure at 0.06 ppm may cut the photosynthesis rate in half. Reliable estimates suggest that 5%–20% of crops such as peanuts, soybeans, and wheat are annually lost due to ozone damage.

**Particulates or particulate matter (PM)** is best discussed by grouping them into categories organized by particle size (radius) or dimensions to evaluate their toxic effects. Particulates are microscopic solid or liquid matter suspended in the atmosphere. The term "aerosol" discussed earlier commonly refers to a particulate/air mixture, as opposed to PM alone. Sources of PM may be natural or man-made. They have an impact on climate and precipitation as well as human health. Depending on the size and nature of the particle, they can also serve as carries of toxic vapors. Ammonia vapors are a prime example of this piggybacking effect.

Subtypes of PM include suspended particulate matter (SPM), respirable suspended particulate matter (RSPM), which generally have a diameter of 10 µm or less, fine particles, which have a diameter between 1 and 2.5 µm, and ultrafine particles, which have a diameter between 0.1 and 1 µm. These are located in the middle, lower portion of Figure 12.4 and are shaded from orange to yellow. Asbestos fibers, which come in several sizes and shapes, are a prime example. The smaller, fine particles (often labeled as **PM 2.5**) and even smaller ultrafine particles are particularly deadly because they are difficult for human lungs to expel through the action of their cilia.

Diesel exhaust fumes and fumes of heavy metals including antimony, beryllium, cadmium, lead, thallium, and vanadium lie in this size range. In addition to posing a hazard due to particle size, each of these has a *permissible exposure limit* or PEL (PELs and similar toxicological terms are discussed in Chapter 15) for an 8-hour work day, since they are carcinogenic. Beryllium, for instance, has a current PEL of 2.0 ug/m$^3$, but based on new health research studies, OSHA is seriously considering lowering this to 0.2 ug/m$^3$.

Once created, dust never dies, though it may disintegrate or aggregate. It just changes location, unless it dissolves in a suitable solvent and reacts with something else.

**Radon, Rn,** is element number 86 and a member of the noble gas family. It might seem, therefore, logical that it is an inert and thus harmless gas. However, though the Rn-222 isotope is naturally occurring and chemically inert, it is nevertheless a radio-active gas produced by the immediate decay of the element radium, itself a decay product originally from uranium (see Chapter 11 for the exact decay scheme). Surface soils and bedrock in many areas of the United States contain sufficient radioactive uranium to create indoor radon problems in buildings and private residences whose construction materials and operating features permit the entry of soil gas. Radon is drawn into a building when air pressure in ground-contact soils is lower than the air pressure in surrounding soils and through cracks or holes in the foundation of the building that allow air movement between the soil and the building interior. Radon radioactivity concentrations in ambient air fluctuate by geography and season, but typical readings of 0.4 pCi/L (pico Curies per liter) are common. Any house has openings, cracks, open windows and doors that allow indoor–outdoor air exchanges, usually at the rate of one exchange every 2–3 hours. Average radon concentrations in the United States, adjusting for seasonal variations, range from 0.4 to 1.5 pCi/L.

Radon also enters through groundwater used as a primary water supply source. When showering, in particular, the radon is exposed as a mist and easily inhaled, putting the decomposing particles in direct contact with the lung tissue, leading to potential cancers. Studies show that indoor radon concentrations fluctuate and that diurnal and seasonal cycles are common, due in part to indoor–outdoor pressure and temperature differences.

Exposure to radon is believed to be the second leading cause of lung cancer, after smoking tobacco. While there is no safe radon concentration in air, the EPA has set a standard or action level of 4 pCi/L.

**Sulfur dioxide**, $SO_2$, is another combustion-related gas. As discussed in a previous section in this chapter, sulfur dioxide is produced when sulfur-containing fuels undergo combustion. It is colorless but can have an odor. Indoor sources include fireplaces, wood stoves, and kerosene space heaters. When inhaled, it contacts and reacts with moisture in the mucous membranes, forming sulfurous acid, which is quite irritating.

Sulfur dioxide is lethal at about 500 ppm, though this concentration is rarely encountered in practical environmental situations. In fact, sulfur dioxide concentrations in the atmosphere rarely exceed 10 ppm. However, long-term, chronic exposure to sulfur dioxide, indoors or outdoors, increases the risk of health problems, particularly for the elderly or chronically ill. Serious medical problems can be made worse by exposure to even low concentrations of sulfur dioxide in indoor air.

Plants are, once again, more sensitive to sulfur dioxide, just they are to ozone. Studies have shown that exposure of some strains of wheat and barley to sulfur dioxide at concentrations of 0.15 ppm for 72 hours can reduce grain yields. Sulfur dioxide kills leaf tissue and bleaches chlorophyll, decreasing the plant's ability to carry out photosynthesis. Soybeans show extensive chlorophyll loss when exposed to sulfur dioxide at 0.8 ppm for just 24 hours. Chronic higher concentrations are sufficient to cause plant death, as evidenced by the loss of plant life around ore refining and smelting operations in such places as Copperhill, TN, and Sudbury, Ontario.

The EPA ambient air quality standards set a limit of 0.14 ppm (365 ug/m$^3$) for a 24-hour period.

# 13 Water Quality and Water Pollution

Water, just to state the obvious, is ubiquitous. Sometimes this is obvious, as in the case of rivers, streams, oceans, or lakes. Sometimes it is less obvious, but still evident, for instance, in the case of rain, irrigation systems, or sewage. Often water hides from sight, but is present in underground aquifers, underground streams, wells, humidity, or even just trapped in clays.

The behavior of water, and materials that are dissolved in or carried by water, can be of significant concern in environmental engineering. Water has a lot of interesting chemical properties, which greatly influence how it needs to be handled in particular situations. In this chapter, what is exciting about water is examined.

## 13.1 CHEMISTRY OF WATER: SUMMARY OF CHEMICAL AND PHYSICAL PROPERTIES

Water is a peculiar material. The molecule $H_2O$ is a very small molecule relatively speaking, but behaves very differently compared to similar-sized molecules. Under conditions encountered on a daily basis, water is usually liquid. Methane, $CH_4$, which is very similar in size and mass, is a gas all the way until it is cooled to $-164°C$ where it liquefies. Water is encountered in all three of its phases, ice, liquid water, and steam, on a daily basis. Most other materials are encountered only in one phase or another. Water also has a distinct property that it does not share with many other compounds, in that the solid phase of water, ice, is less dense than the liquid phase, allowing ice to float. Water is also often referred to as the universal solvent as it is capable of dissolving a lot of different types of compounds, although "universal" is a bit of an exaggeration.

Most of these properties arise from the structure of the molecule, which gives rise to a particular type of interaction between two or more water molecules, called hydrogen bonding or H-bonding. It is important to distinguish H-bonding from other types of bonds, as this is a bond that is between molecules rather than one that is within a molecule such as a covalent bond.

H-bonds occur between water molecules due to the fact the oxygen atoms are highly electronegative. They attract electrons significantly more easily than hydrogen atoms do. While the electrons are still shared in the covalent bonds of water, they are not shared evenly. This leads to two consequences. One, the oxygen has extra electron density; this can be imagined as if it was an extra inflated balloon of electrons. Two, the hydrogen is particularly deprived of electrons, since the sharing is uneven. Although the bond between the oxygen and the hydrogen is still strong, the hydrogen atoms could still potentially accommodate more electrons near it due to this uneven distribution. This uneven distribution also leads to the area near the oxygen atoms in the water molecule to become partially negatively charged and the areas near the hydrogen atoms to become partially positively charged. See Figure 13.1.

Now that what is happening inside a water molecule has been clarified, what happens when two (or more) water molecules encounter each other can be considered. Since the oxygen on each water molecule has an abundance of electrons, the electron-deficient hydrogen atoms on other water molecules are going to be attracted to them. When the oxygen on one water molecule encounters the hydrogen of another water molecule, a strong interaction occurs, as the hydrogen draws off some electron density from the oxygen. This interaction is what is referred to as H-bonding. H-bonding leads to the particularly interesting behavior of water molecules, as they are strongly attracted to each other. See Figure 13.2.

The expansion of water when freezing is also attributed to H-bonding interactions. Because of how electrons tend to arrange themselves inside an atom, water molecules have a bent shape, with electrons arranged in a tetrahedral shape within the atom. As water molecules attach to each other to form ice, the fundamental shape of the water molecule, combined with the linear nature of H-bonds, governs the macroscopic arrangement. The network formed has gaps in between the water molecules as the ice structure forms. These little gaps lead to a decrease in density, as they are empty and have no mass in them, and each water molecule can be thought of as taking up more space than before the water froze, as shown in Figure 13.3 (more space, same mass, reduced density).

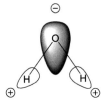

**FIGURE 13.1** Polarity of water molecules. Electrons spend more time in proximity to the oxygen atom, building up a negative charge near it, and positive charges near the hydrogens.

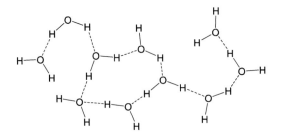

**FIGURE 13.2** H-bonding in liquid water. Solid lines indicate covalent bonds; dotted lines indicate H-bonds between water molecules.

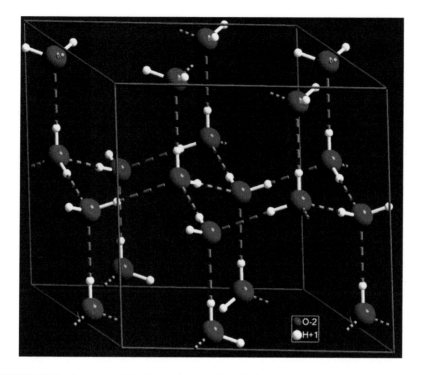

**FIGURE 13.3**  Structure of ice. Dotted connections indicate H-bonds. Image created by and generously released into the public domain by Wikipedia user Materialscientist.

From these discussions, it can be seen that the structure of water will continue to be relevant in its behavior.

## 13.2  SOLUTES, SOLVENTS, AND SOLUTIONS

Many fluids encountered on a daily basis are not pure materials, but a combination of various materials referred to as a solution. As presented in Chapter 3.4, solutions are homogeneous mixtures; that is to say that it is not possible to separate the components of the solution by filtration. A common example would be white wine or coffee. Once any coffee grounds are filtered out, the remaining liquid is predominantly water, but caffeine molecules, sugar molecules (assuming one takes coffee with sugar), and a variety of other flavorful chemicals are all floating around in the water. Although not enjoyable like coffee is, groundwater, wastewater, water runoff, and many other materials encountered in an engineering setting are also solutions.

A solution is composed of a solvent and one or more solutes. The solvent is whatever medium makes up the majority component of the material at hand. Here, the discussion will focus on solutions where this medium is a liquid, although solid and gas solutions also exist. In many cases this liquid is water, so it will generally be assumed that this is the case, although exceptions do arise. Solutes are those materials that are in lesser amounts within the solution. Solutes are said to be dissolved in the solvent.

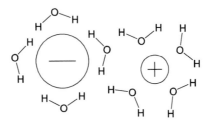

**FIGURE 13.4**   Solvated ions. The negatively charged chlorine is surrounded by the positive hydrogen atoms on the water molecules. The positively charged ion is surrounded by the negatively charged oxygen side of the water molecules.

Dissolution generally occurs because the solute and solvent particles involved in the process have favorable interactions. This gives rise to the rule about solubility that is "like dissolves like." For instance, water and ethanol readily dissolve in each other, as both are small molecules, and have OH groups leading to H-bonding. Because of this similarity, the two materials attract each other strongly and can mix easily. Oil and water on the other hand do not share this similarity. The molecules that make up oil tend to be long chains of carbons and hydrogens, which cannot interact with water molecules nearly as strongly; thus water molecules stay with the water molecules, oil molecules stick with each other, and the two materials stay separated.

Salts are often soluble in water. Salts are composed of anions and cations, which tend to separate from each other in solution. Water molecules can interact with both cations and anions favorably due to the fact that the oxygen side is negatively charged and the hydrogen side is positively charged. These charges allow water molecules to surround anions and cations that are entering the solution with opposite charges, making the overall interaction favorable. See Figure 13.4.

An important aspect of this process is how many particles enter solution. This sounds deceptively simple, but does get somewhat complicated, as some materials generate multiple particles per formula unit in solution. For example in the case above, sodium chloride, NaCl, table salt, would generate two particles in solution per "NaCl" dissolved; one cation, $Na^+$, and one anion, $Cl^-$. The van't Hoff factor, $i$, of a solute is equal to the number of particles generated in solution. For example, for NaCl, $i = 2$.

For ionic compounds, the van't Hoff factor is generally equal to the number of ions in the compound. In particularly concentrated solutions, the real value of $i$ tends to be lower than the theoretical value, but this is usually not a concern. For molecular compounds, $i$ is usually equal to 1.

## 13.3   COLLOIDS

Mixtures between particles of various sizes form different types of media. In solutions, the particles are sufficiently small that the medium may be called homogeneous or uniform throughout. Particles in solution do not settle out. In suspensions, the medium is not homogeneous, that is, there are large regions of one material or another, and they settle out over time. For example, if muddy river water is left to sit still, the mud will settle to the bottom. As initially presented in Chapter 3.10, in between solutions and suspensions, is another type of mixture known as a colloid, where the particles in solution are large enough that the medium is not really homogeneous, but the particles also do not settle out. Colloids in water solution are formed when the surface of a large particle is able to interact with water favorably. This condition can come about in several different ways.

Hydrophilic colloidal particles are large molecules that have surface regions that can interact with water. Certain proteins and polymers have many polar components, and these polar surface components can form dipole–dipole interactions and even H-bonds with the water medium. This allows for these particles to persist suspended in water as a colloid.

Colloids can also form from hydrophobic materials as long as the surface of the hydrophobic material becomes hydrophilic somehow. There are several ways this can happen. Ions can settle on the surface of the colloid from within the solution, stabilizing the hydrophobic particle in solution. Surfactants, or soaps, are molecules designed or chosen to specifically accomplish this end. These are molecules that have a large hydrophobic group on one end of the molecule and a hydrophilic group on the other end. The hydrophobic tails of these soap molecules bury themselves in the hydrophobic material, but the hydrophilic end stays outside, on the surface. This process allows for the formation of small droplets of otherwise water-insoluble materials to be coated with a water-soluble surface. These droplets can then be suspended in water solution in a colloid, not settling out of the medium. This interaction can be seen in Figure 13.5.

As colloids can often allow for the persistence of waste materials in water supplies, an important consideration for colloids is how to force a colloid to decompose

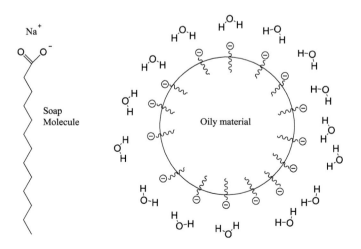

**FIGURE 13.5**  Soap and colloids. Left: A typical soap molecule is a long hydrocarbon chain with a carboxylate end. Right: The long hydrocarbon chain dissolves into oily material, while the carboxylate at the end remains at the surface of the globule, interacting with the water medium.

into two distinct phases; that is to force the suspended material out of solution. The stability of a colloid relies on the droplet of the suspended phase being prevented from colliding with each other and coagulating into larger and larger droplets, which would eventually separate from the solution. The droplets tend to not coagulate due to the surface charges introduced by the surfactant, which repels individual droplets from each other. Introducing excess charged particles can reduce the effectiveness of this repulsion. Heating can also allow for more energetic collisions, overcoming the repulsive forces of the surfactant and allowing the suspended colloidal particles to coagulate. Higher technology processes, like filtration through a semipermeable membrane, can also break up colloids.

Two commonly used chemicals in wastewater purification are alum, which is aluminum sulfate, and iron(III) chloride. These compounds are used in a process called flocculation, wherein colloidal particles are removed from wastewater by forcing them to destabilize and precipitate. There are different mechanisms by which this process works. At low concentrations of additives, the aluminum or iron ions neutralize the surface ions of colloidal particles, allow them to collide, form particles large enough to precipitate, and settle out. At higher concentrations, the action is believed to proceed by the formation of sweepfloc. Sweepfloc is formed when aluminum hydroxide or iron hydroxide particles precipitate after the addition of alum or iron(III) chloride. The aluminum or iron hydroxide precipitates form spontaneously and can capture colloidal contaminants by entrapping them as the precipitate is formed, or by collisions, since the hydroxide precipitates do not have charged surfaces to repel colloidal particles. In either case, the chemical additives destabilize the colloid allowing for its removal from wastewater streams.

The intrepid reader may choose to perform a small experiment at home to observe the phenomenon. Milk, in fact, is a colloidal solution, with casein molecules keeping butterfat globules suspended in solution. With the addition of an acid, say lemon juice or vinegar, the milk can be curdled. In the curdling process, the acid added neutralizes the charges on the casein, allowing the fat molecules to collide and form large clumps. These clumps quickly separate from the solution rising to the top.

## 13.4 COMMON SOURCES OF CONTAMINATION OF WATER SUPPLIES

There is a wide array of potential sources for chemical contamination of water supplies, often ones easily overlooked. Some of these sources are nearly unavoidable, but many sources can best be controlled at the point of release. Awareness of these sources should help significantly in avoiding water contamination.

Automotive care and related activities can generate a significant amount of waste. Oil spills can easily be washed into local water supplies as part of water runoff. Small amounts of oil can cause much larger volumes of water to become polluted. Oil slicks can cover very large surface areas, causing various problems, from leaching toxic compounds into the water, to cutting off the oxygen influx from the air into the body of water. This can have a major impact on wildlife.

Heavy metal and other inorganic waste can be released by industrial activities and urban traffic, both incidentally and particularly when their disposal is poorly regulated. For instance, in normal operation, cars can release a number of metallic contaminants in the form of small particulate dust from wear and tear. These metals, such as copper and zinc, can enter water supplies as runoff. Industrial processes can generate even worse metallic contamination in the form of mercury and lead waste. These metals tend to be quite toxic (see Chapter 6).

Inorganic wastes such as sulfur, nitrates, and phosphates are also released as part of industrial waste. Phosphates are also present in some detergents used in washing machines and dishwashers. Nitrates and phosphates in particular have an impact on the biosphere as they can act as nutrient sources. Algal blooms can be triggered by the introduction of phosphate or nitrate into a waterway.

Mining activities can be devastating to local environments. As one example, gold mining tends to rely on very toxic chemicals in ore processing. In the past, mercury was used to dissolve gold out of gold-rich gravel. Sometimes more chemically aggressive techniques are required, which still use chemicals like sodium cyanide, which are exceptionally toxic. Inevitably these compounds find their way into local water supplies and tend to be very damaging to ecosystems.

## 13.5   COD AND BOD

Chemical oxygen demand (COD) and biochemical oxygen demand (BOD) are two values used to quantify the amount of organic compounds present in water. Usually defined in mg $O_2$/L water, it more specifically measures the amount of oxygen required to completely decompose all organic molecules in a sample of water to $CO_2$, $H_2O$, and, with the frequent occurrence of ammines in organic compounds, $NH_3$. COD is tested chemically, whereas BOD specifically looks at the behavior of organisms in the water sample with relation to consumption of organic molecules.

### 13.5.1   COD Measurements

Measuring COD is done through the chemical oxidation of the materials in a water sample. It is quantified as milligrams of oxygen consumed per liter of water sample. Most often this determination is made using potassium dichromate. The dichromate ion is a very strong oxidizer and will consume all present organic material, transferring its oxygen atoms to them. By measuring how much dichromate reacts with a given sample, the amount of organic material can be quantified. The unbalanced equation for the reaction is shown in Equation 13.1.

$$C_xH_yO_z + Cr_2O_7^{2-} \rightarrow CO_2 + H_2O + Cr^{3+} \tag{13.1}$$

The technique used for this determination is back titration. In this technique, a known amount of excess dichromate is added to the water sample to be analyzed and allowed to completely react. This leaves some of the dichromate unreacted. Another compound that also reacts with the dichromate is then added in a controlled manner, until the dichromate is completely consumed. To detect the completion of the reaction, an indicator compound is usually used, which changes color when the dichromate is completely reacted.

**Example 13.1:** Calculating Chemical Oxygen Demand (COD)

A stream behind an automotive repair shop is suspected to be contaminated with oil runoff. A sample of stream water is taken, and the COD is determined by the following means. 10 mL of .25 M potassium dichromate is added to 100 mL of stream water. The reaction is stirred and allowed to react. The solution is then titrated with ammonium iron(II) sulfate ($(NH_4)_2FeSO_4$, FAS (commonly termed ferrous ammonium sulfate), and 14.0 mL of 1.0 M FAS is required to completely react with the solution. What is the chemical oxygen demand of this water sample?

*Solution*

The technique described above is called back titration. Initially, an excess amount of potassium dichromate is added to the water sample and is allowed to react with any organic compounds. Once this reaction is over, another compound, FAS, is added in a controlled manner until all of the remaining potassium dichromate is consumed. The calculations have to start from this second reaction. FAS can donate one electron per molecule of FAS, and dichromate has to be reduced from $Cr^{6+}$ to $Cr^{3+}$. So that gives

$$Cr_2O_7^{2-} + 6FAS \rightarrow 2Cr^{3+} + byproducts \tag{13.2}$$

Based on this, the information from the titration can be used to find the amount of dichromate left over.

$$0.014 \text{ mL FAS} \times 1.0 \text{ M FAS} = 0.014 \text{ mol FAS} \tag{13.3}$$

$$0.014 \text{ mL FAS} \times \frac{1 \text{ mol } Cr_2O_7^{2-}}{6 \text{ mol FAS}} = 0.00233 \text{ mol } Cr_2O_7^{2-} \tag{13.4}$$

Since it is now known that 0.00233 mol of $Cr_2O_7^{2-}$ remained following the reduction of the organics, the amount of dichromate used to consume the organic materials in the water sample can be back calculated.

$$0.0025 \text{ mol } Cr_2O_7^{2-} - 0.00233 \text{ mol } Cr_2O_7^{2-} = 0.00017 \text{ mol } Cr_2O_7^{2-} \tag{13.5}$$

$$0.00017 \text{ mol } Cr_2O_7^{2-} \times \frac{7 \text{ oxygen}}{2 Cr_2O_7^{2-}} = 0.000595 \text{ mol Oxygen} \tag{13.6}$$

$$0.00595 \text{ mol Oxygen} \times \frac{16 \text{ g Oxygen}}{1 \text{ mol Oxygen}} = 0.00952 \text{ g Oxygen}$$

$$= 9.5 \text{ mg Oxygen} \tag{13.7}$$

Using stoichiometry, the amount of dichromate that reacted was calculated, from which the oxygen demand was determined by calculating the amount of oxygen that the dichromate carried with it. This gives 9.5 mg of O, in 100 mL of sample.

$$\frac{9.5 \, mg}{100 \, mL} \times \frac{1000 \, mL}{L} = \frac{95 \, mg}{L} \qquad (13.8)$$

The chemical oxygen demand of the given sample is 95 mg/L.

## 13.5.2 BIOCHEMICAL OXYGEN DEMAND

BOD is a measure of the amount of oxygen that microbial life consumes in the process of digesting organic contaminants in a sample of water. COD measures the amount of all organic material, whereas BOD is specific to organic compounds that are biodegradable. Since natural digestion processes are significantly slower than the dichromate reaction used to measure COD, BOD is usually measured after a 5-day incubation at 20°C.

BOD generally relies on the measure of dissolved oxygen present in solution, and a variety of techniques exist to measure this value. Dissolved oxygen sensors exist, so the value can be measured most easily with instrumentation. There are also digestion bottles that have a special cap designed to sense pressure changes, and as oxygen is consumed, the pressure change can be translated into a concentration change allowing for the display of a BOD value. Regardless of the exact procedure, BOD is determined by measuring the initial amount of oxygen present in a reactor and remeasuring the value after incubation. Like COD, BOD can be expressed as a mg/L value representing mg of oxygen consumed per liter of sample water in 5 days.

## 13.6 CARBON DIOXIDE POLLUTION

Carbon dioxide is of significant environmental concern for a large number of reasons, but here the issues related to carbon dioxide and water are addressed. Carbon dioxide is not only emitted by natural processes of life but also is emitted in significant amounts by industrial activities conducted by humans. Transportation, energy generation, and even other industrial processes like smelting of iron or calcination in the production of concrete are all significant sources of anthropogenic carbon dioxide and are discussed in more detail in Chapter 2.

Carbon dioxide enters the atmosphere, but a significant portion of it dissolves into the ocean and other bodies of water. The dissolution of carbon dioxide in ocean water leads to a chemical equilibrium between $CO_2$ and carbonic acid:

$$CO_2 + H_2O \rightleftharpoons H_2CO_3 \qquad (13.9)$$

Carbonic acid, $H_2CO_3$, undergoes further equilibrium reactions that generate $H_3O^+$, acidifying the water

$$H_2CO_3 + H_2O \rightleftharpoons H_3O^+ + HCO_3^- \qquad (13.10)$$

$$HCO_3^- + H_2O \rightleftharpoons H_3O^+ + CO_3^{2-} \qquad (13.11)$$

The industrial age began about 250 years ago, and fortunately ocean acidity records have been kept for that entire span of time. In 250 years, the surface pH of oceans has decreased by roughly 0.1 on the pH scale, roughly from 8.2 to 8.1. This looks deceptively small, but the pH scale is a logarithmic scale, and this corresponds to just under a 30% increase in $H_3O^+$ concentration.

The impact on the biology of ocean creatures is complex, but this change, especially as it continues, will be disruptive to the ocean biome. The metabolism and biology of many animals are beyond the scope of this text, but one particular problem is very much rooted in chemistry and is likely to have a very large impact. This is the impact on shelled creatures that heavily rely on calcium in their life cycle.

The problem lies in Le Chatelier's Principle and the complex equilibria in $CO_2$ dissolution and the formation of solid $CaCO_3$. The math is complex, but these equilibria lead to a decrease in the concentration of $CO_3^{2-}$ in the ocean, in preference to the formation of $HCO_3^-$. The problem is that this impacts the solubility of $CaCO_3$ as that reaction has the following equilibrium:

$$CaCO_3(s) \rightleftharpoons Ca^{2+} + CO_3^{2-} \qquad (13.12)$$

$$K = \left[Ca^{2+}\right]\left[CO_3^{2-}\right] \qquad (13.13)$$

As the $CO_3^{2-}$ concentration is reduced, the $Ca^{2+}$ concentration must increase for equilibrium to be maintained. This change, however, means that $Ca^{2+}$ is generally more soluble in water, reducing the ability of shelled animals to use $Ca^{2+}$ in building their shells. Changes to ocean water acidity can have significant impact on coral, shellfish, and crustaceans whose shells are effectively being dissolved by the excess $CO_2$ being introduced into their environment.

## 13.7    ENVIRONMENTAL FATE AND TRANSPORT OF SELECTED WATER POLLUTANTS

There can be significant concern over the fate of chemical contaminants that reach a body of water. Contaminated surface water can adversely affect local wildlife and can potentially enter the food chain leading up to human consumption. Groundwater contamination can also be a significant problem, as underground transport of contaminants may carry contaminants a significant distance from their point of origin. Some contaminants released into the air (see the above section for $CO_2$ impact) also have a significant impact on water resources.

### 13.7.1    DDT

DDT, or dichlorodiphenyltrichloroethane, is a pesticide that was used broadly in the middle of the last century. DDT was eventually found to be acutely and chronically toxic to humans and wildlife and was eventually banned globally in the Stockholm Protocol signed in 2001. DDT is a good example of why the chemical properties of some compounds can make them particularly hazardous. DDT is very chemically stable, and its primary decomposition products, DDE (dichlorodiphenylchloroethylene) and DDD (dichlorodiphenyldichloroethane) are similarly toxic. None of these organic molecules are water-soluble. This means that the molecules, once sprayed, have a tendency to not move within the environment. They would not be washed into soil by water, and although they could move within a waterway if the compound was directly introduced into the water, it would settle into sediments or evaporate to return to the environment within weeks. Depending on the source, the environmental lifetime of DDT has been estimated at anywhere between 2 and 30 years, which likely depends very strongly on local environments.

Of course, DDT has a number of proponents, as it is highly effective at eradicating mosquito populations. This is of particular interest in areas with high incidence of malaria, as controlling mosquito populations can significantly reduce infection rates. Balancing the medical needs of these populations against the broader health of the environment requires significant consideration. Other examples of pesticides and their toxicities are discussed in more detail in Chapter 15.6.

### 13.7.2 PHARMACEUTICAL POLLUTION

Many types of pharmaceuticals have an impact on the environment. Antihistamines that pass into waterways have been shown to disrupt biofilms, which are a crucial component of aquatic biomes. Ethinylestradiol is a hormone that is a component of certain types of birth control pills. Released into the environment, they have induced changes in fish populations damaging the reproductive capabilities of the males of some species. High concentrations of antibiotics are also present in some bodies of water, increasing the probability of the emergence of antibiotic-resistant infectious bacteria.

Pharmaceuticals are a problem in part because many drugs are not altered chemically in the body. They act on the body for a time before they are removed by the body as waste, unchanged, and are flushed down the drain. Similarly, many cosmetics meet the same fate, as they are washed off in a shower at the end of a day. Although the nature of the effect of these trace compounds is only partially understood, some studies have shown them to have an impact on wildlife living in waterways affected by these waste streams. Unfortunately most of these compounds are present in low concentrations in wastewater. Treatment plants are not generally equipped to handle the removal of these contaminants. For this reason, many municipalities have drug recovery programs, so that unused doses of medication can be returned to pharmacies and properly disposed rather than flushed down drains into waterways.

### 13.7.3 DIOXINS

Dioxins are a class of compounds that are **persistent organic pollutants (POP)**, which are generally formed as a by-product of industrial processes or waste incineration. Dioxins include three subclasses of compounds, polychlorinated dibenzo-p-dioxins, polychlorinated dibenzofurans, and polychlorinated biphenyls (PCBs). The basic structures are shown in Figure 13.6, with a varying number and location of chlorines making up the family of compounds. This is a fairly broad group of chemicals, and behaviors vary across different molecules within the group, but the more dangerous members of the family are highly toxic.

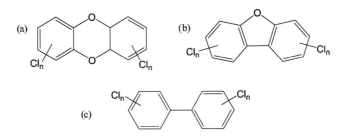

**FIGURE 13.6**   Dioxins: (a) polychlorinated dibenzo-p-dioxins, (b) polychlorinated dibenzo-furans, and (c) polychlorinated biphenyls.

Dioxins are poorly water-soluble, but dissolve easily in fat. These properties reduce the likelihood of dioxins being spread through environmental routes. They are, however, readily absorbed by animal and plant life. Since dioxins are also chemically very stable, they tend to persist in the organisms that absorb them. This leads to a process called **bioaccumulation or biomagnification**. As animals higher in the food chain eat large amounts of contaminated animals lower in the food chain, the animals higher up build up a higher concentration of dioxins. This accumulated concentration is then passed on to animals even higher in the food chain, increasing the concentration of dioxins further. This increases the toxic impact of dioxins the further up the food chain the compound is passed. Dioxins were unwanted by-products or contaminants in the manufacturing of Agent Orange, an herbicide used as a defoliant in the Vietnam War. They are known teratogens. Dioxins are discussed again in Chapter 15.

Production of dioxins was banned in 2001 by the Stockholm Convention, and generation as a by-product of waste disposal has been significantly curtailed by improvements in incineration technology. Nonetheless, dioxin is a persistent pollutant, and care must be exercised in areas with potential sources of exposure. Dioxins are discussed in more detail in Section 15.5.

Other examples of emerging contaminants and toxicants, which have environmental fate and consequence, are presented and discussed in Chapter 15, Sections 15.6–15.9. They include atrazine, arsenic, PFAS, and 1,4 dioxane as well as some other toxicants.

# 14 The Chemistry of Hazardous Materials

## 14.1 BACKGROUND

By the very nature of certain elements, compounds, and mixtures thereof, the field of chemistry necessarily deals with hazardous materials. A hazardous material may be naturally occurring or synthetic but, regardless, it is characterized in one of three ways: toxic to human health; capable of undergoing spontaneous or uncontrolled chemical reaction, including combustion or explosion; or causing irreversible harm to the environment. To put it more succinctly, the three areas of concern associated with hazardous materials are toxicity, flammability, and reactivity. The subject of chemical toxicity is presented in some detail in Chapter 15. Flammability is discussed in more detail in Section 14.4 of this chapter. Reactivity is a complex subject and is discussed on a broad basis in Sections 14.1–14.3. Corrosion is an important factor in assessing hazardous materials and is normally a topic treated under reactivity and so will be discussed in that context. While biohazards, electronic, and radioactive wastes are three other categories of hazardous materials, they will not be discussed here. In industry, some substances used as reactants may be hazardous while others may be generated as by-products during the synthesis and manufacturing of a desired material or product. The field of green chemistry, discussed in Section 14.6, seeks to minimize both the use of starting materials and the generation of products, whether intended or unintended by-products or contaminants, that are hazardous materials, by seeking alternate reaction routes for synthesis of the end product. The objective is to stop environmental pollution and exposure to hazardous materials by eliminating them at the source, an essential theme in green chemistry.

Nevertheless, as a result, governments of many modern countries have developed a system of regulations to protect the general population, its workers, consumers, and the environment at large. These regulations or laws are designed to reduce or minimize exposure to hazardous materials, protect people and the environment. The vast majority of environmental protection legislation in the United States was passed beginning only in the 1970s with the births of the Environmental Protection Agency (EPA) and the Occupational Health and Safety Administration (OSHA) in 1970. In addition, the US federal government also established the Department of Transportation (DOT) in 1966, which is also concerned with a similar mission of protecting the health and safety of transportation systems and infrastructure.

Major federal legislation protecting environmental quality include The Clean Harbors Act of 1899,[1] The Clean Air Act of 1970, The Water Pollution Control

---

[1] The very first environmental legislation enacted in the United States designed to protect the Mississippi River and still enforced today.

Act of 1972, The Safe Drinking Water Act of 1974, The Resources Conservation and Recovery Act of 1976 (RCRA), and The Toxic Substances Control Act of 1976 (TSCA). The reader will note that most of these laws have undergone substantive revisions and updates resulting in subsequent amendments such as The Comprehensive Environmental Response, Compensation, and Liability Act (CERCLA) of 1980, The Superfund Amendments and Reauthorization Act (SARA) of 1986, The Clean Air Act Amendments (CAAA) of 1990, and the recent overhaul of the 1976 TSCA signed into law in June of 2016 (known formally as The Frank R. Lautenberg Chemical Safety for 21st Century Act), giving the EPA authority to request safety data on chemical substances that are both new and already in use and in the marketplace. All are examples of legislation overseen and enforced by the EPA and are designed to protect the health and safety of the general public.

Meanwhile, The Laboratory Standards Act of 1991 and The Hazard Communication Standard of 1983, a.k.a. "Employee Right-to-Know Law," and the origin of the Material Safety and Data Sheet (MSDS), now simply called the Safety Data Sheet (SDS), are examples of laws overseen and enforced by OSHA. They are designed to protect the health and safety of individual workers in the workplace. An example of an SDS is given in Appendix B, while an example of a National Fire Protection Association (NFPA) warning label is given in Appendix C. It should be noted that OSHA estimates that more than 43 million workers produce or handle hazardous chemicals in more than 5 million workplaces in the United States. For a more comprehensive review and understanding of applicable laws and regulations, what they mean, and division of responsibility, the interested reader should consult other relevant references on the subject.

Two quick notes of enlightenment the authors would like to offer are as follows. First, authority and responsibility to regulate health and safety matters are divided up among various federal agencies (over which there may be overlap). OSHA has jurisdiction to oversee health and safety questions occurring within the borders or "inside the fence" of any industrial or commercial place of business or employment. An employer has the responsibility and duty to provide and ensure safe and healthy working conditions for all of its employees. Violations should be brought to the attention of OSHA, headquartered in the US Department of Labor. EPA has jurisdiction to protect all elements of the environment—air, water, land, or areas "outside of the fence," which affect the health and safety of the general population and citizenry at large. In short, EPA has the authority and responsibility to protect environmental quality against contamination or pollution. DOT has the authority and responsibility to ensure the safe transport of all hazardous materials whether by air, land, or waterway, i.e., anything in motion. The Coast Guard and US Army Corps of Engineers share responsibility for environmental protection and transportation safety within and adjacent to all navigable waterways of the United States and all their tributaries, in essence covering all waterways of the United States.

Second, a distinction exists between *recommending* entities or agencies and *enforcement* or *regulatory* agencies. For example, all three of the aforementioned agencies are regulatory and enforcement agencies, capable of conducting investigations and assessing penalties or fines. On the other hand, organizations such as the American National Standards Institute (ANSI), the NFPA, and National Institutes of Occupational Safety and Health (NIOSH) are recommending bodies. (Note that ANSI and NFPA are private organizations while NIOSH and the Agency for Toxic Substances and Disease Prevention (ATSDR) are each government agencies operating under the Center for Disease Control and Prevention (CDC), which, in turn, is part of the Department of Health and Human Services.) For example, while ANSI recommends that eyewash fountains should be inspected and flushed once per week, it has no authority to enforce this or issue fines. In contrast, OSHA regulations require only annual testing of this facility, but can issue fines for noncompliance.

NIOSH has researched and compiled a list of *recommended exposure levels* or RELs, which recommend maximum concentrations of exposure to chemical substances encountered in the workplace. The American Conference of Government and Industrial Hygienists (ACGIH), still another private organization, based on its own, independent research, also recommends *threshold limit values* or TLVs (see Chapter 15 for definitions and explanation of this and other toxicological terms used in this paragraph), based on *time-weighted averages* or TWAs of a 40-hour work week. But OSHA is free to adopt and enforce standards from any combination of RELs and TLVs it deems appropriate and can stipulate and enforce its own set of exposure concentrations, called *permissible exposure levels* or PELs. The Chemical Safety and Hazard Investigation Board (CSB), established in 2005, is separate government agency, which investigates chemical accidents and makes recommendation to reduce or prevent future accidents through OSHA, EPA, and DOT, but does not have authority to assess penalties or fines. Further, the Consumer Product Safety Commission (CPSC), which oversees enforcement of the Federal Hazardous Substance Act (FHSA), established in 1960, has enforcement power for all *consumer products*. It is clear that there are at least five federal agencies and several private organizations that conduct risk assessment for chemicals and at least three federal entities with enforcement power.

## 14.2 THE CHEMISTRY OF FOUR COMMON ELEMENTS

Some elements more than others seem to turn up in the chemical formulas of hazardous substance. Of particular note are hydrocarbons and organic substances (separately discussed in Chapters 8 and 9) containing oxygen, sulfur, and chlorine. The chemistry of each of these elements is discussed below.

### 14.2.1 HYDROGEN

Hydrogen ranks ninth in overall abundance by mass on Earth (on a combined basis of crust, oceans, and atmosphere) at 0.9%. In its elemental state, which is rare, it exists as a gaseous diatomic molecule ($H_2$). Since it is a very reactive element, it is found in its free state only in trace amounts on the Earth. It is much more commonly found in compounds such as water, acids, and hydrocarbons. However, on the scale of the universe, it is the most abundant element, accounting for over 70% of the mass of the entire universe.

Elemental hydrogen is an odorless, colorless, tasteless gas at room temperature. It is nontoxic. Hydrogen can be liquefied when compressed under a pressure greater than 294 psi and a temperature less than $-234.5°C$ ($-390°F$). It has a wide flammability range, with lower to upper explosive limits of 4%–74% v/v. See Section 6.5 for an explanation of upper and lower explosive limits.

Hydrogen may be produced by several methods. In the first method, steam is passed over hot coals (coke) under high temperature and pressure. This process, often referred to as coal gasification, produces a mixture of carbon monoxide, CO, and hydrogen, $H_2$, gases, often called "water gas," as follows:

$$C(S) + H_2O(g) \rightarrow CO(g) + H_2(g) \qquad (14.1)$$

The carbon monoxide is converted to carbon dioxide by passing the water gas with additional steam over a catalyst such as iron (III) oxide. The carbon dioxide is subsequently removed by passing it through an alkaline solution (a neutralization reaction), leaving pure hydrogen gas behind.

A second method of generating hydrogen involves reacting steam with methane, $CH_4$, at high temperatures, as follows:

$$CH_4(g) + H_2O(g) \rightarrow 3H_2(g) + CO(g) \qquad (14.2)$$

The "water gas" thus produced is treated by the method described above to remove the carbon monoxide, thereby isolating free hydrogen gas.

A third method for producing hydrogen gas involves a complex, multistep process, which can be summarized by the reaction between propane, $C_3H_8$, and water, as follows:

$$C_3H_8(g) + 6H_2O(g) \rightarrow 3CO_2(g) + 10H_2(g) \qquad (14.3)$$

A fourth method involves the reaction of steam with methanol vapor, $CH_3OH$, using a zinc oxide or copper oxide catalyst, as follows:

$$CH_3OH(g) + H_2O(g) \rightarrow CO_2(g) + 3H_2(g) \qquad (14.4)$$

As described earlier for the process in reaction 4.1, the carbon monoxide is converted to carbon dioxide, which is removed by passing it through an alkaline solution. Suffice it to say that water at a temperature of $700°C–1000°C$ in the absence of a catalyst in contact with a hydrocarbon has a better than average chance of producing hydrogen gas.

Other reactions that produce hydrogen involve select metals with an acid. These reactions also fall under the heading of corrosive reactions or simply corrosion. These are really oxidation–reduction reactions. For example, consider the reaction between tin, Sn, or aluminum, Al, with hydrochloric or sulfuric acid, respectively.

$$Sn(s) + 2HCl(aq) \rightarrow SnCl_2(aq) + H_2(g) \qquad (14.5)$$

$$Al(s) + 3H_2SO_4(aq) \rightarrow Al_2(SO_4)_3(aq) + 3H_2(g) \qquad (14.6)$$

Hydrogenation is a term used when converting polyunsaturated fats or oils to saturated ones. It refers to the addition of hydrogen atoms to carbon–carbon double or triple bonds to convert it into a single bond. The example of converting hexane to hexene is given below, but the interested reader should consult Chapters 8 and 9 of this book for further examples of saturation of alkenes and alkynes.

$$C_6H_{12}(l) + H_2(g) \rightarrow C_6H_{14}(l) \tag{14.7}$$

Hydrogen and oxygen react in more than one ratio to form either water or hydrogen peroxide, (illustrating the **Law of Multiple Proportions**), as shown below:

$$2H_2(g) + O_2(g) \rightarrow H_2O(l) \tag{14.8}$$

$$H_2(g) + O_2(g) \rightarrow H_2O_2(l) \tag{14.9}$$

Hydrogen peroxide, with a Lewis structure of H—O—O—H, is classified as a strong oxidizing agent, depending on its concentration. Concentrations made for consumer consumption start at 1%–3%, when used for topical applications as antiseptics for minor cuts and wounds. It may also be used as a sterilizing agent and disinfectant to kill *E. coli* bacteria, botulism, and salmonella, as well as other microorganisms that cause disease. Hydrogen peroxide is also available at 6% and is used to bleach hair. It oxidizes the pigment melanin to colorless products.

Hydrogen peroxide is also available at concentrations ranging from 30% to 90% for industrial applications. The 30% solution is used to bleach cotton, wool, straw, paper, and leather. The 70% solutions are used in the chemical industry to initiate red-ox reactions. The 90% solution is used in the aerospace industry as an auxiliary booster in propulsion.

## 14.2.2  OXYGEN

At 49.5%, oxygen ranks as the most abundant element on the Earth by mass, though the vast majority of this oxygen is found in a bound state to other molecules. It is found in its free state in air, which is 21% oxygen by number or volume (and 78% nitrogen). The rest of the oxygen is found in compounds such as water, oxides of metals, and mixed or hydrated salts found in the oceans, crust, and atmosphere of the Earth, and, of course, in all living plants and animals as part of the biosphere. Like hydrogen, it exists as a diatomic molecule in its free or natural state, two oxygen atoms connected by a double bond. Oxygen is one of the seven diatomic molecules in its natural state, the other six being $H_2$, $N_2$, $F_2$, $Cl_2$, $Br_2$, and $I_2$. It is also one of 11 elements that exist as gases at room temperature and pressure. The other ten are hydrogen, nitrogen, fluorine, chlorine, helium, neon, argon, krypton, xenon, and radon.

Oxygen forms compounds with both metals, e.g., calcium oxide and iron (III) oxide, and with nonmetals, e.g., sulfur dioxide and nitrous oxide. It forms many oxo-anions, such as carbonates, nitrates, and silicates, as well as alumino-silicates, all of which comprise much of the Earth's crust (lithosphere). In addition, it is the base atom of many of the functional groups in organic compounds, e.g., alcohols, ethers, ketones, carboxylic acids, etc. At 3.44, it ranks second only to fluorine (3.98) in electronegativity.

As is well known, oxygen is needed for human survival. People breathe in oxygen and exhale carbon dioxide. This process is known as respiration. Although atmospheric oxygen is constantly consumed by all living zoological organisms, it is replenished during photosynthesis, in which plants convert carbon dioxide and water into oxygen and carbohydrates, in the presence of sunlight and chlorophyll. Scientists now calculate that at least 50%–60% of the atmospheric oxygen in the world is generated by the life processes of marine phytoplankton, one-celled plants that thrive in the oceans. At least another 20%–25% are estimated to come from the Amazonian rainforest in South America. These land and ocean plant masses coexist and maintain the concentration of oxygen in the atmosphere at about 21%.

At ambient temperatures, oxygen is a colorless, odorless, tasteless gas. At 20°C and 1.0 atmosphere pressure, it has density of $1.429 \times 10^{-3}$ g/cm³, slightly denser than dry air at $1.204 \times 10^{-3}$ g/cm³. It liquefies at −183°C (−297°F). As a gas, oxygen ranks third behind fluorine and chlorine in oxidizing potential, so it is a powerful oxidizing agent in its natural state. Its common oxidation state is −2. It must be present to support combustion of hydrocarbon fuels as well as the oxidation of iron to form rust.

Ozone is an allotrope of oxygen and is present and desirable in the stratosphere, though not in the troposphere, e.g., as a component of smog (see Chapter 12). The Lewis structure of ozone is

$$O = O - O \Leftrightarrow O - O = O$$

It is well known that chlorinated hydrocarbons from aerosol spray have been responsible for the destruction of the ozone layer in the stratosphere. The interested reader should consult the appropriate section of Chapter 12 for the sequence of reactions.

Oxygen is ubiquitous and forms many naturally occurring and commercially important **oxo-anions**. Examples include nitrate, $NO_3^-$ (found in ammonium nitrate, $NH_4NO_3$, used both as a fertilizer and an explosive), permanganate, $MnO_4^-$ (found in potassium permanganate, $KMnO_4$, used to treat bacterial or fungal infections of the skin, as well as a general disinfectant to treat wastewater), and sulfate, $SO_4^{2-}$ (found in calcium sulfate, $CaSO_4$, used to make plaster).

Two compounds of particular interest containing oxo-anions are potassium chromate, $K_2CrO_4$, and potassium dichromate, $K_2Cr_2O_7$. Chromium is present in the hexavalent oxidation state, $Cr^{6+}$, in both the chromate anion, $CrO_4^{2-}$, and the dichromate anion, $Cr_2O_7^{2-}$. Potassium chromate can be produced first by treating chromite ore, represented as $FeCr_2O_4(s)$, with potassium carbonate and atmospheric oxygen, i.e., air in a kiln. Potassium dichromate can be subsequently produced by treating potassium chromate with sulfuric acid as follows:

$$2\,K_2\,CrO_4(s) + H_2SO_4(aq) \rightarrow K_2Cr_2O_7(aq) + H_2O(l) + K_2SO_4(aq) \quad (14.10)$$

Potassium chromate (a yellow solid) may be produced from potassium dichromate (an orange solid), and vice versa, by altering pH conditions as follows:

$$K_2Cr_2O_7(aq) + 2KOH(aq) \rightarrow 2K_2CrO_4(aq) + H_2O(l) \qquad (14.11)$$

$$2K_2CrO_4(aq) + HCl(aq) \rightarrow K_2Cr_2O_7(aq) + H_2O(l) + 2KCl(aq) \qquad (14.12)$$

Both of these compounds have many commercial applications. They have been used as simple pigments in paints. Because of their strong oxidizing power, they have also been used in the manufacturing of textile dyes, the preservation of wood and leather, and chrome electroplating baths.

Prior to the 1990s, **hexavalent chromium compounds** were employed as wastewater treatment agents to prevent the formation of mineral scale that would otherwise form in the circulating waters used in air conditioners and industrial cooling towers. The presence of this scale buildup reduces the efficiency by which heat is transferred to water. Cooling towers are large, tall structures employed in industries such as petroleum refineries, chemical plants, and power plants. They are also used where large-scale air-conditioning or refrigeration systems are needed, such as at large hotels and convention centers. The wastewater generated by these systems was kept on-site in outdoor, unlined surface pools. Rainwater would naturally cause the hexavalent chromium ions to migrate, then seep into and contaminate the groundwater. Although trivalent chromium is an essential trace element for human health, hexavalent chromium is toxic and potentially carcinogenic, depending on the exposure route, concentration, and time. In the 2000 movie "Erin Brockovich," it was alleged that long-term exposure to and consumption of water contaminated with hexavalent chromium caused cancer and a myriad of other health problems. In reality, the evidence now is much stronger than the evidence was back in 2000. EPA has set a maximum contaminant level (MCL) of 100 ug/L for drinking water. EPA has also used TSCA to prohibit the use of water treatment agents containing hexavalent chromium in HVAC systems.

An important cautionary note about one class of organic compounds containing oxygen bears mentions, i.e., ethers, already mentioned in Chapter 9. Ethers are a class of compounds that contain the $R_1$-O-$R_2$ linkage or functional group, where $R_1$ and $R_2$ are alkyl or aryl groups (hydrocarbon chains). Diethyl ether, $C_2H_5$-O-$C_2H_5$, is an example of this functional group. Here, $R_1 = R_2$, but that is not always the case. Tetrahydrofuran is another example of an ether. Ethers stored in glass bottles or other suitable containers for long periods of time may leak, if not sealed tightly, absorb and react with ambient oxygen, and form unstable organic peroxides. **Organic peroxides** have a $R_1$-O-O-$R_2$ linkage and may be sensitive to mechanical shock and explode. If there are signs of crystallization or even cloudiness present, great care must be exercised in moving or opening such a bottle or container.

## 14.2.3 CHLORINE

Chlorine is a member of the halogen family and like oxygen is a diatomic gas at room temperature and pressure. It is a yellowish-green, sharp-smelling gas, discovered by Carl Scheele in 1774. It has the highest electron affinity, ranks third highest in electronegativity (behind fluorine and oxygen), and ranks second only to fluorine in oxidizing potential. It is found in nature in the form of dissolved salts in seawater and solid deposits in salt mines. Chlorine is produced from the electrolysis of aqueous sodium chloride and is one of the top ten chemical feedstocks manufactured in the United States for commercial uses.

Chlorine once had a reputation for use in many popular commercial products that have been long since banned because of health hazards, including carbon tetra-chloride (a nonflammable degreasing agent used in the dry cleaning industry), DDT or dichloro-diphenyl-tetrachloroethane (a powerful insecticide), and PCBs or poly-chlorinated biphenyls (insulating fluids used in capacitors, transformers, and other electrical equipment, plastics)

Inhalation of chlorine gas is its principal toxic risk. Initial exposure to chlorine gas (at concentrations of about 20 ppm) causes dizziness, headache, nausea, and severe inflammation to the eyes, nose, and throat. Prolonged exposure (>1 hour) at moderate concentrations (>34 ppm) may cause congestion of the lungs, leading to pulmonary edema. The production of hydrogen chloride gas, $HCl(g)$, which later can be dissolved in water to make the commonly used laboratory reagent, hydrochloric acid, $HCl(aq)$, is as follows:

$$H_2(g) + Cl_2(g) \rightarrow 2HCl(g) \qquad (14.13)$$

When chlorine atoms are incorporated into a compound, such as water, the process is known as **chlorination**. Chlorine gas may be dissolved directly in water to simultane-ously produce hypochlorous acid, $HOCl(aq)$, and hydrochloric acid, $HCl(aq)$. Chlorine atoms can also be introduced by adding sodium hypochlorite (a.k.a. common bleach), $NaClO$, a liquid, or calcium hypochlorite, $Ca(ClO)_2$, a solid, to water. This process, referred to as the **chlorination of water**, is a standard method for treating and disin-fecting drinking water. It is also used to treat wastewater for the removal of certain organic and inorganic compounds, e.g., amines and nitrogenous substances. First,

$$Cl_2(g) + H_2O(l) \rightarrow HClO(aq) + HCl(aq) \qquad (14.14)$$

or

$$NaClO(l) + H_2O(l) \rightarrow HClO(aq) + NaOH(aq) \qquad (14.15)$$

Note that reaction (14.14) produces an acidic solution, while reaction (14.15) produces an alkaline one. The pH of either reaction can be adjusted as desired with appropriate reagents.

Hypochlorous acid, $HOCl(aq)$, then reacts with ammonia, $NH_3$, (emanating from amines or nitrogenous substances) to form chloramine (a.k.a. monochloramine), $NH_2Cl(g)$:

$$HOCl(aq) + NH_3(g) \rightarrow NH_2Cl(g) + H_2O(l) \qquad (14.16)$$

Subsequent treatment of chloramine (often used as a disinfectant by itself) with addi-tional HOCl produces dichloramine, $NHCl_2$, and trichloramine, $NCl_3$, which are responsible for swimming pool smell.

$$HOCl(aq) + NH_2Cl(g) \rightarrow NHCl_2(g) + H_2O(l) \qquad (14.17)$$

Chlorine and ammonia may also react directly to form compounds, with or without the presence of water, which are particularly hazardous to health. Gaseous ammonia and chlorine can react to produce ammonium chloride, $NH_4Cl$, and chloramine, $NH_2Cl$. Regardless of the pathway, $NH_2Cl$ is a very toxic gas, which can lead to suffocation or asphyxia. Di- and trichloroamines, $NHCl_2$ and $NCl_3$, may also be formed by autocatalytic reaction or consecutive reactions and are equally hazardous.

$$2NH_3(g) + Cl_2(g) \leftrightarrow NH_2Cl(g) + NH_4Cl(s) \qquad (14.18)$$

$$2NH_3(g) + Cl_2(g) + H_2O(l) \rightarrow NH_4ClO(s) + NH_4Cl(s) \qquad (14.19)$$

Ammonium hypochlorite, $NH_4ClO$, shown in reaction (14.19), is a white powder soluble in water. It is similar to sodium hypochlorite, common bleach, and a strong oxidizing agent, and it is corrosive to human tissue. In short, mixing ammonia-based cleaning solutions with chlorine-based bleaches will most likely produce a variety of toxic substances and should be avoided.

Chlorine dioxide, $ClO_2$, is a strong oxidizing agent. It is one of three formally recognized oxides of chlorine and is probably the one of the greatest commercial importance. The other two are sodium chlorite, $NaClO_2$, and sodium hypochlorite, $NaClO$. At room temperature, $ClO_2$ is a red-yellow gas and quite unstable. The oxidation state of Cl in this compound is +4. It can spontaneously decompose into its elements as follows:

$$2ClO_2(g) \rightarrow Cl_2(g) + 2O_2(g) \qquad (14.20)$$

## 14.2.4 SULFUR

Sulfur is one of the few elements that actually exists as a free solid in nature. It is a brittle yellow solid. Deposits have been found in Mexico, Italy, and Japan, mostly near the edges of hot springs or volcanoes. It constitutes about 0.06% by mass of the crust. It is in the same family as oxygen, i.e., the chalcogen group. Sulfur exists in two chief allotropic forms: orthorhombic and monoclinic sulfur, both atomically structured as $S_8$ molecules. The melting point of sulfur varies between 113°C and 119°C, depending on the allotrope and its crystalline structure. When elemental sulfur is first ignited, it melts into a liquid, $S_8(l)$, which burns with a blue flame, before vaporizing. The combustion with oxygen produces sulfur dioxide, $SO_2(g)$, a yellowish gas with a characteristic suffocating odor.

Approximately 90% of all the sulfur produced throughout the world is burned to produce sulfur dioxide gas, $SO_2(g)$. Most of this amount is used to manufacture sulfuric acid, $H_2SO_4$, and most of this—about 35 million tons—is produced in the United States. Some $SO_2$ is used as a food preservative, in dried fruits and wines, since it is especially toxic to molds and bacteria. It is also used as bleach for textiles and wood pulp used in papermaking.

Sulfuric acid, once known as the oil of vitriol, is the least expensive of the commercially prepared acids and can be prepared and shipped in pure form. It is used by the fertilizer industry to convert insoluble phosphate rock into soluble superphosphate, which supplies phosphate ions required by growing plants. Sulfuric acid is also used as the electrolytic solution in automobile batteries, as well as in metal treatment known as a "pickling process." Pickling here means dipping the metal part in a bath of sulfuric acid to remove rust and other contaminants.

As discussed in Chapter 4, sulfuric acid is a substantial component of acid rain. Coal and oil contain sulfur deposits, which release sulfur into the atmosphere when burned. Sulfur dioxide and sulfur trioxide form and react with moisture to produce sulfuric (and sulfurous) acid. The interested reader should consult the appropriate sections of Chapter 4 for further clarification of this problem.

Elemental sulfur also reacts with most metals. For example, if a mixture of iron and mercury is heated in the presence of sulfur, the metals unite to form mercury (II) sulfide and iron (II) as shown below:

$$8Hg(1) + S_8(1) \rightarrow 8HgS(s) \qquad (14.21)$$

$$8Fe(1) + S_8(1) \rightarrow 8FeS(s) \qquad (14.22)$$

Another sulfur compound of note is hydrogen sulfide, $H_2S$. It is a gas that is characterized by the smell of rotten eggs. The origin of this description is the actual smell of rotten eggs that comes from hydrogen sulfide, which forms by the bacterial decomposition of sulfur-containing proteins in the yokes of eggs. It is an extremely toxic gas, roughly equivalent in toxicity to hydrogen cyanide gas. It damages the human body by attacking the respiratory enzymes, where even trace amounts, i.e., 1–10 ppm, can cause headaches and nausea. "Prudent Practices" lists the odor threshold detection limit for $H_2S$ as between 0.001 and 0.1 ppm with the qualifier that "olfactory fatigue occurs quickly at higher concentrations" (Prudent Practices, 1995, pg. 342). This means that an individual exposed to concentrations of 1.0 ppm or higher experiences progressively faster desensitizing effects, such that by the time the exposure level reaches 10 ppm, which OSHA lists as its PEL-TWA, the individual may be unable to detect the odor and be unaware that he has already reached the action level.

## 14.3    THE CHEMISTRY OF SOME CORROSIVE MATERIALS

### 14.3.1    THE NATURE OF CORROSIVITY

The Consumer Protection and Safety Commission (CPSC) defines a **corrosive substance** as any consumer product that destroys any living tissue such as the skin or eyes by chemical reaction. The DOT defines **corrosive material** as a liquid or solid that causes full-thickness destruction of human skin at the site of contact, or a liquid that reacts with aluminum or steel surfaces at a rate exceeding 6.25 mm/year at a test temperature of 54°C (130°F).

The OSHA Laboratory Standard defines a corrosive substance as a substance with a pH less than 2.0 or greater than 12. These are generally substances that would irreversibly damage living human tissue by chemical reaction. By this definition, an acidic solution such as 0.010 M HCl (pH = 2.0) or alkaline solution such as 0.010 M NaOH (pH = 12.0) would be just at the initial limits of being corrosive, while solutions of 0.0010M HCl (pH = 3.0) and 0.0010 M NaOH (pH = 11.0) would not. Thus, whether a substance is corrosive depends on both the nature of the substance and its aqueous concentration. Primary classes of corrosive substances include acids, bases, and acidic and basic anhydrides. However, many other products whose pH falls outside the range of 2.0–12 would also be considered corrosive.

## 14.3.2   THE NATURE AND PROPERTIES OF ACIDS AND BASES

The basic concept of acids and bases, as well as the distinction between strong versus weak, has already been introduced and discussed in Chapter 1. Recall that a **strong acid** is one that dissociates and ionizes completely or 100% into hydrogen ions and its conjugate base, while a **weak acid** is one that only partially dissociates and ionizes, i.e., less than 10% and typically less than 5%. This means then that for every 100 molecules of acid present initially, five molecules or less actually separate into hydrogen ions, $H^+$, the active acid species. Thus, 95% or more remain as undissociated parent or source molecules, essentially inactive. The same distinction holds for strong and weak bases, except that the ions here are hydroxide ions, $OH^-$, and their conjugate acid.

In general, acids have the properties of being strong dehydrating agents, neutralizing bases, reacting with metals, producing potentially large amounts of heat, turning litmus dye paper to a red color, and tasting sour. Examples of acid–base neutralization reactions are given in Chapter 1.

There are four, commonly encountered, strong acids:

Hydrochloric Acid—HCl
Nitric Acid—$HNO_3$
Sulfuric Acid—$H_2SO_4$
Perchloric Acid—$HClO_4$

Hydroiodic, HI, acid and hydrobromic acid, HBr, are also strong acids but are rarely encountered. These acids are strong because they dissociate and ionize completely into hydrogen ions and their respective conjugate bases. HCl, $HNO_3$, and $HClO_4$ are **monoprotic** acids since they produce a single hydrogen ion, $H^+$, per parent molecule, when dissociated and ionized. $H_2SO_4$ is a **diprotic** acid since it produces two hydrogen ions, $H^+$. Those that produce more than one hydrogen ion are collectively referred to as **polyprotic** acids.

In addition, nitric, sulfuric, and perchloric acids are oxidizing acids, due to the positive oxidation states of nitrogen, sulfur, and chlorine in the respective acids. This means that each acid is a powerful oxidizing agent and capable of initiating a red-ox reaction. This property makes them particularly hazardous and dangerous to work with.

Representative examples of reactions involving these four acids are as follows:

$$Mg(s) + 2HCl(aq) \rightarrow MgCl_2(aq) + H_2(g) \tag{14.23}$$

$$FeO(s) + H_2SO_4(aq) \rightarrow FeSO_4(aq) + H_2O(l) \qquad (14.24)$$

$$Al_2O_3(s) + 6HNO_3(aq) \rightarrow 2Al(NO_3)_3(aq) + 3H_2O(l) \qquad (14.25)$$

$$ZnCO_3(s) + H_2SO_4(aq) \rightarrow ZnSO_4(aq) + CO_2(g) + H_2O(l) \qquad (14.26)$$

$$Cu(s) + H_2SO_4(aq)(conc) \rightarrow CuSO_4(aq) + SO_2(g) + H_2O(g) \qquad (14.27)$$

$$C(s) + H_2SO_4(aq)(conc) \rightarrow CO_2(g) + 2SO_2(g) + 2H_2O(g) \qquad (14.28)$$

The following three reactions involve the same two reactants, $HNO_3(aq)$ and $Zn(s)$, yet generate different products. This sequence illustrates the variety of products possible, depending on the concentration of nitric acid as well as the relative stoichiometric ratios of the two reactants, i.e., zinc to nitric acid. Two other reactions (not shown here) generating different products are also possible.

$$5Zn(s) + 12HNO_3(aq) \rightarrow 5Zn(NO_3)_2(aq) + 6H_2O(l) + N_2(g) \qquad (14.29)$$

$$3Zn(s) + 8HNO_3(aq) \rightarrow 3Zn(NO_3)_2(aq) + 4H_2O(l) + NO(g) \qquad (14.30)$$

$$Zn(s) + 4HNO_3(aq)(conc) \rightarrow Zn(NO_3)_2(aq) + 2H_2O(l) + NO_2(g) \qquad (14.31)$$

While the list of weak acids is much longer, running into several hundred, a few examples of commonly encountered weak acids include the following:

Acetic Acid—$CH_3COOH$
Phosphoric Acid—$H_3PHO_4$
Hydrofluoric Acid—$HF$
Nitrous Acid—$HNO_2$
Sulfurous Acid—$H_2SO_3$

Weak acids have $K_a$ values, while strong acids do not, because their $K_a$ values approach infinity. Chapter 4 has a table of an abridged list of $K_a$ values. See the CRC Press' *Handbook of Chemistry and Physics* for a comprehensive list of commonly encountered weak acids and their respective $K_a$ values. Generally speaking, the smaller the $K_a$ value, the smaller the percent ionization and the weaker the acid.

### 14.3.3   SELECT WEAK ACIDS OF INTEREST

Acetic acid is found in many food products in the concentration range of 3%–6% by mass and is responsible for the sour taste. Vinegar is about 5% acetic acid by mass. Concentrated acetic acid, 80%–99% by mass, is known as glacial acetic acid. It is produced through the reaction of methanol and carbon monoxide. It releases a vapor, which, if inhaled, produces a choking and suffocating sensation and can readily damage the bronchial tract and eyes. It is highly corrosive. It is an OSHA category 3 flammable liquid and an NFPA class II combustible liquid (see Section 4.4 for clarification). OSHA's PEL for the vapor concentration is 10 ppm averaged over an 8-hour workday.

Phosphoric acid, $H_3PO_4$, is the most commonly encountered mineral acid. It has three hydrogen ions to give for every molecule of parent acid. Each of the three dissociation and ionization steps is progressively weaker. Though a weak acid, it is one of the stronger weak acids. Phosphoric acid is most familiar to the general public as a component of household cleaning products like Lime-Away and removes rust and calcium-based deposits. It is found in many popular carbonated beverages. In the chemical industry, it is used to manufacture organophosphate pesticides as well as to produce superphosphate fertilizers. Finally, it is used to produce mono-ammonium phosphate, the dry chemical fire extinguisher.

Hydrofluoric acid is a solution of hydrogen fluoride gas in water, HF(aq). While it is weak, it is a deceptively dangerous acid. It is the only acid that *cannot* be stored in a glass container. It attacks glass, concrete, and natural carbonaceous materials such as wood, leather, and rubber. Used in cleaning metals, it is also used as an etching agent in glass manufacturing and in the electronics industry. It etches glass by reacting with silicon dioxide, $SiO_2$, to form gaseous or water-soluble silicon fluorides as follows:

$$SiO_2(s) + 4HF(aq) \rightarrow SiF_4(g) + 2H_2O(l)$$

$$SiO_2(s) + 6HF(aq) \rightarrow H_2SiF_6(aq) + 2H_2O(l)$$

It is also a precursor to almost all fluorine compounds, including pharmaceuticals such as fluoxetine (Prozac) and polytetrafluoroethylene (Teflon).

If spilled on the skin, it will penetrate the epidermis in a matter of a few minutes. Unlike other acids, it is dangerous in two ways. Because of its small size, undissociated HF molecules can readily penetrate the skin, creating no immediate symptoms. Once inside, dissociated, free hydrogen ions are a powerful dehydrating and corrosive agent and will burn the skin tissue. The fluoride ions are sufficiently small and particularly mobile and will simultaneously seek out unbound calcium and magnesium ions in blood as well as in surrounding tissue and bone, precipitating them as fluoride salts. Dilute HF solutions penetrate deeply before dissociating in their respective ions, thus delaying injury and the onset of symptoms. Severe burns occur after exposure to concentrated solutions (50% or greater) of HF to 1% or more of the body surface area. Furthermore, by interfering with the calcium metabolism, it may cause systemic toxicity and eventual cardiac arrest and fatality, after contact with as little as 25 square inches (160 cm²) of skin.

Hydrofluoric acid solutions have an acrid smell and generate a vapor that can cause severe burns to the eyes. At concentrations of 10–15 ppm, HF vapor is irritating to the eyes, skin, and respiratory tract. Brief exposure (5 minutes) of 50–250 ppm may be fatal.

Exposure to HF acid or its vapors requires immediate medical assistance. Depending upon the type and nature of exposure, as well as the concentration of the

HF acid, flushing the affected area(s) with copious amounts of water is recommended (15 minutes or longer), followed by treatment with calcium gluconate topical gel/ointment or injections by trained professionals.

### 14.3.4 SELECT BASES OF INTEREST

Bases, on the other hand, yield hydroxide ions when dissolved in water. They react with fats and greases, including skin oils, neutralize acids, and potentially produce large amounts of heat, feel slippery and taste bitter, and turn litmus dye paper to a blue color. Examples of acid–base neutralization reactions are provided in Chapter 1 of this book.

Commonly encountered strong bases include the following:

**Sodium Hydroxide, NaOH**, is a constituent in Drano, Liquid-Plumr, and Easy-Off Oven Cleaner used to unclog drains or dissolve grease and fat from inside ovens. Sodium hydroxide solutions are also used at wastewater treatment facilities to isolate metallic compounds as water-soluble metallic hydroxides. It is an industrial reagent used in the purification of petroleum products, reclaiming of rubber, and the processing of textiles and paper. It is also a raw material used in the manufacturing of soap, rayon, and cellophane.

**Potassium Hydroxide, KOH**, is principally used in the chemical industry in the production of soft soaps, fertilizers, pharmaceutical products. It is also used as an electrolyte in alkaline batteries.

**Calcium Hydroxide, Ca(OH)₂**, is chiefly used as a raw material in the production of plaster, mortar, and cement. It is also used an effective neutralizing agent for miscellaneous acid spills by emergency responders.

**Magnesium Hydroxide, Mg(OH)₂**, also known as milk of magnesium, is the active ingredient in many over-the-counter stomach ant-acid. Though a strong base, it has limited solubility, which tempers its alkalinity and makes it safe for ingestion.

Representative examples of reactions of strong bases with metals are as follows:

$$2\,Al(s) + 6\,NaOH(aq) \rightarrow 2\,Na_3AlO_3(aq) + 3\,H_2(g) \qquad (14.32)$$

$$Pb(s) + 2\,KOH(aq) \rightarrow K_2PbO_2(aq) + H_2(g) \qquad (14.33)$$

A few examples of weak bases include the following:

Ammonium Hydroxide—$NH_4OH$
Chloramine (a.k.a. Mono-Chloramine)—$NH_2Cl$

Like weak acids, weak bases have $K_b$ values, as a measure of their potential strength. See Chapter 4 for a table of an abridged list of $K_b$ values or consult the CRC Press' *Handbook of Chemistry and Physics* for a more comprehensive list.

### 14.3.5 ANHYDRIDES OF ACIDS AND BASES

When a metallic or nonmetallic element combines with oxygen, the resulting compound is either a metallic oxide or a nonmetallic oxide, respectively. Potassium and calcium are examples of two metals that can burn in oxygen to form metallic oxides, i.e., potassium oxide, $K_2O(s)$, and calcium oxide, $CaO(s)$.

Each of these compounds can in turn react with water to form a base, as follows:

$$K_2O(s) + H_2O(l) \rightarrow KOH(aq) \tag{14.34}$$

$$CaO(s) + H_2O(l) \rightarrow Ca(OH)_2(aq) \tag{14.35}$$

The term anhydride refers to a substance from which water has been removed. In this case, $K_2O$ and $CaO$, which are metallic oxides, are referred to as **basic anhydrides**. Since both oxygen and water are ubiquitous substances in the environment, these reactions may occur or be in progress at any time.

Similarly, when a nonmetallic element, such as sulfur or carbon, combines with oxygen, the resulting compound is a nonmetallic oxide, i.e., sulfur dioxide, $SO_2$, or carbon dioxide, $CO_2$. Each of these compounds can in turn react with water to form an acid, as follows:

$$SO_2(g) + H_2O(l) \rightarrow H_2SO_3(aq) \tag{14.36}$$

$$CO_2(g) + H_2O(l) \rightarrow H_2CO_3(aq) \tag{14.37}$$

Thus, sulfur dioxide, sulfur trioxide, and carbon dioxide are **acidic anhydrides**.

The upshot of this is to recognize that metallic oxides behave as bases while nonmetallic oxides behave as acids. The chief reason that the pH of rainwater is 5.6, and not 7.0, is due to the presence of carbon dioxide in the atmosphere. Further, as discussed in Chapter 12, one of the main sources of acid rain is the generation of sulfur dioxide (and sulfur trioxide), pollutants that emanate from coal burning plants. Depending upon the ready availability of moisture in the atmosphere, these can in turn lead to the formation of sulfuric acid and sulfurous acid (illustrated in Equation 4.36) and lower the pH to even 4.0.

## 14.4 THE CHEMISTRY OF PYROPHORIC SUBSTANCES

A **water-reactive substance** is an element or compound that reacts with water. The compound(s) produced are a flammable gas that ignites spontaneously, or a corrosive, or a toxic compound harmful to human health. In a sense, it is a kind of hydrolysis, but with products that meet at least one of the three previously mentioned criteria.

An **air-reactive substance**, for which the proper technical term is **pyrophoric substance**, is an element or compound that spontaneously ignites when exposed to oxygen and moisture (water vapor), which is normally present in ambient air. Normal temperature and pressure (NTP) are assumed, i.e., 25°C and 1.0 atmosphere pressure.

There are several categories of substances that meet either or both of these criteria. The first alkali metals in elemental form—lithium, sodium, potassium, cesium, and rubidium—are soft, silvery metals. They can be cut with a knife. All are water-reactive, producing hydrogen gas (flammable) and an alkaline solution (corrosive). Elemental sodium reacts with water as follows:

$$2\,Na(s) + 2H_2O(l) \rightarrow 2NaOH(aq) + H_2(g) \qquad (14.38)$$

Further, they react with oxygen to form metal oxides, peroxides (only sodium and potassium), and a superoxide (only potassium), as follows:

$$4Li(s) + O_2(g) \rightarrow 2Li_2O(s) \qquad (14.39)$$

$$4Na(s) + O_2(g) \rightarrow Na_2O(s) \qquad (14.40)$$

$$2\,Na(s) + O_2(g) \rightarrow Na_2O_2(s) \qquad (14.41)$$

$$K(s) + O_2(g) \rightarrow KO_2(s) \qquad (14.42)$$

Mixtures of the oxides and peroxides are possible and common for sodium, and mixtures of oxides, peroxides, and superoxides are possible and common for potassium, when they are burned in air. A superoxide is more properly called a hyperoxide and is a highly reactive oxidizing agent.

Other metallic salts, such as iron (III) chloride, $FeCl_3$, can react with water as follows:

$$FeCl_3(s) + 3H_2O(l) \rightarrow Fe(OH)_3(s) + 3HCl(aq) \qquad (14.43)$$

Since iron (III) hydroxide is insoluble and precipitates, and HCl remains, the solution is potentially acidic and hence, corrosive.

Ionic hydrides of metals, such as sodium hydride, are also water-reactive, producing hydrogen gas, which is flammable, and an alkaline solution, which is corrosive. Sodium hydride is just one example.

$$NaH(s) + H_2O(l) \rightarrow NaOH(aq) + H_2(g) \qquad (14.44)$$

A short list of common water reactive and pyrophoric substances is given in Table 14.1.

---

**TABLE 14.1**
**Water-Reactive and Pyrophoric Substances**

| Alkali Metals | Pyrophoric; Water-Reactive |
|---|---|
| Nonmetal hydrides | Pyrophoric |
| Metal powders (finely divided) | Pyrophoric |
| Lithium hydride | Water-reactive |
| Calcium carbide | Water-reactive |
| Picrate salts | Explosive |
| Organic lithium reagents | Pyrophoric |
| Grignard reagents | Pyrophoric |

---

## 14.5   THE CHEMISTRY OF FLAMMABLE SUBSTANCES

The flammability of a substance is one of three important characteristics in determining the hazardous properties of a substance. The flammability of a substance is largely determined by the amount of vapor it has to produce to ignite and burn, when a suitable source of ignition is present. The necessary amount of vapor, in turn, is generally determined by surrounding temperature, called the **flashpoint** (occasionally and mistakenly referred to as the ignition point). The flashpoint of a flammable liquid, then, is the minimum temperature at which it gives off vapor to form an ignitable mixture with air across the surface of the liquid or within a test vessel. For example, the flash point of toluene is 39°F (4.0°C). This means that this is the minimum temperature necessary for toluene to be ignited and sustain or propagate a flame, given the presence of a suitable source of ignition. Below this temperature, it will not burn.

The mixture with air is a critical key. Every substance has a unique minimum and maximum gas or vapor concentration below which and above which it will not burn. These minimum and maximum concentrations are expressed in percent by volume and are called **lower flammable limit** and **upper flammable limit**, respectively. They may be also referred to as **upper and lower explosive limits**. Flammable or explosive limits are always measured in ambient air, which is 21% oxygen and 78% nitrogen. The upper and lower limits change when the composition of air changes. For example, the flammability range of toluene is 1.1%–7.1% by volume. This means that toluene will burn only when its vapor concentration in air is between 1.1% and 7.1% and a suitable source of ignition is present.

Two other terms deserve mention—**fire point** and **autoignition point**. The fire point is the temperature at which a substance gives off sufficient vapor such that it continues to burn once the ignition source is removed. It is the temperature at which there is self-sustained combustion. The fire point is usually 30°F–50°F (17°C–72°C) higher than the flash point of a substance.

The autoignition point is the minimum temperature at which the confined vapor of a flammable liquid is sufficient to self-sustain combustion *without* the presence of a suitable source of ignition. The autoignition point for toluene, for example, is 896°F (480°C). Flashpoints, fire points, and autoignition points for common, flammable liquids have been compiled in the scientific and engineering literature and are available on Internet resources.

There are at least four different ways or approaches to classifying flammable substances, individually developed by different government agencies or private organizations: OSHA, NFPA, CSPC, and DOT. As one might guess, there is considerable overlap among the first two. The first two systems are the most extensive and will be presented here in bulleted format.

### 14.5.1   OSHA CLASSIFICATION SYSTEM

OSHA defines a **flammable liquid** as any liquid having a vapor pressure not exceeding 40 psi at 100°F (37.8°C) and having a flashpoint at or below 199.4°F (≤93°C). OSHA further defines four categories of flammable liquids in Table 14.2.

**TABLE 14.2**
**OSHA Categories of Flammable Liquids**

- Category 1 Flammable Liquid: a liquid with a flashpoint below 73.4°F (<23°C) and an initial boiling point at or below 95°F (≤35°C)
- Category 2 Flammable Liquid: a liquid with a flashpoint below 73.4°F (<23°C) and an initial boiling point at or above 95°F (≥35°C)
- Category 3 Flammable Liquid: a liquid with a flashpoint at or above 73.4°F (≥23°C) but at or below 104°F (≤60°C).
- Category 4 Flammable Liquid: a liquid with a flashpoint above 140°F (>60°C) but below 199.4°F (≤93°C).

## 14.5.2 NFPA CLASSIFICATION SYSTEM

NFPA defines **flammable liquid** as any liquid that has a flashpoint below 100°F (<37.8°C) and a boiling point below 100°F (<37.8°C). NFPA further recognizes three classes of flammable liquids in Table 14.3.

NFPA further defines a **combustible liquid** as any liquid having a flashpoint at or above 100°F (≥37.8°C). NFPA recognizes three classes of combustible liquids in Table 14.4.

**TABLE 14.3**
**NFPA Categories of Flammable Liquids**

- Class 1A Flammable Liquid: a liquid with a flashpoint below 73°F (<22.8°C) and a boiling point below 100°F (<37.8°C).
- Class 1B Flammable Liquid: a liquid with a flashpoint below 73°F (<22.8°C) and a boiling point at or above 100°F (≥37.8°C).
- Class 1C Flammable Liquid: a liquid with a flashpoint at or above 73°F (>22.8°C) but below 100°F (<37.8°C).

**TABLE 14.4**
**NFPA Categories of Combustible Liquids**

- Class II Combustible Liquid: a liquid with a flashpoint at or above 100°F (≥37.8°C) but below 140°F (<60°C), other than a mixture having 99% or more of the volume of its components with flashpoints equal to or greater than 200°F (≥93.3°C). An example of a class II combustible liquid is acetic acid.
- Class IIIA Combustible Liquid: a liquid with a flashpoint at or above 140°F (≥60°C), but below 200°F (≤93.3°C). An example of a class IIIA combustible liquid is creosote oil.
- Class IIIB Combustible Liquid: a liquid with a flashpoint at or above 200°F (≥93.3°C). An example of a class IIIB combustible liquid is ethylene glycol.

## 14.6 THE CHEMISTRY OF EXPLOSIVES

An explosive is a chemical or nuclear material that can undergo a sudden and rapid self-propagating decomposition reaction, which results in the release of heat, sound, light, and sudden pressure changes. Only chemical explosives will be addressed in this section.

Chemical explosives are generally encountered as integral parts of any of the following:

- Ammunition used in hunting or sporting events.
- Artillery and munitions used in wartime.
- Industrial strength charges used to mine coal, drill for oil wells, tunneling through rock during construction of transportation systems.

There are two general types of explosives:

- Explosives substances that are unique compounds, which readily detonate or can be made to detonate easily.
- Explosive articles, which are manufactured to contain explosive substances such as cartridges and military bombs.

The decomposition of an explosive is referred to as a **detonation**. Generally speaking, a single substance is detonated to form one or more decomposition products. Consider the case of nitroglycerin, for example, whose molecular structure is shown in Figure 14.1. It is a nitrate ester with three nitro groups, ($NO_2$), experiencing steric repulsion between neighboring oxygen atoms and between neighboring nitrogen atoms, providing a meta-stable state.

Now note that four different products result from its decomposition—$CO_2$, $N_2$, $O_2$, and $H_2O$—given in Equation 14.42. (Though not shown in the reaction, carbon monoxide gas, CO, and carbon particulates (soot) may also be produced, depending on actual reaction conditions.)

$$4C_3H_5N_3O_9(l) \rightarrow 6N_2(g) + 12CO_2(g) + 10H_2O(g) + O_2(g) \qquad (14.45)$$

Note that four moles of nitroglycerin undergo decomposition to form 29 moles of product. Further note that the reactant is in the liquid state, while all of the products are in the gaseous state, requiring a massive increase in space or volume. Taken

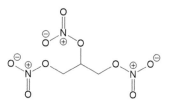

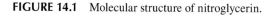

**FIGURE 14.1** Molecular structure of nitroglycerin.

together, these facts provide a preliminary explanation of why a sudden pressure change might take place in a confined space.

However, the complete explanation is more complex. A detonation is a sudden ignition that is accompanied by an energy wave through the body of an explosive. The speed of this wave is called the **detonation velocity**. Most explosive substances are exceptionally powerful not just because of the large amount of energy released, but also because of the exceptionally high rate at which this energy propagates. The average detonation velocity is 23,000 ft/s (7000 m/s).

The gases and vapors in this reaction are heated to 5400°F (3000°C) by the heat energy released. The increase in temperature causes the substances to expand so rapidly that the passage of an accompanying shock wave exerts an exceedingly high pressure on the nearby surroundings. Unlike combustion, the detonation does not require the presence of atmospheric oxygen. Instead, detonation involves the rapid decomposition of a single substance that simultaneously acts as both an oxidizing and a reducing agent. At elevated temperatures accompanying the detonation, the heated gases expand to occupy approximately 10,000 times their initial volume in less than a millionth of a second. This prodigious expansion produces not only a shock wave but is also accompanied by luminescence and an extraordinarily loud sound.

An explosive detonates at a unique temperature, called **detonation temperature**. This property is difficult to measure since it depends on the explosive's density, geometric shape, and other factors. It is roughly equal to its melting point.

It is interesting to note that pure nitroglycerin, a nitrate ester, is a thick, oily liquid whose physical appearance resembles water. However, commercial-grade nitroglycerin is generally a viscous, pale yellow liquid that is extremely sensitive to spontaneous decomposition. Dropping a container on a hard surface or even a slight jarring of it may trigger its detonation. Given this sensitivity to decompose, pure nitroglycerin is not safe to transport and is impractical for general use as an explosive. Nitroglycerin also hydrolyzes when exposed to moisture in the air, producing a mixture of nitric acid and glycerol and becoming even more sensitive to spontaneous decomposition. Unfortunately, terrorists in the Middle East misused this property of nitroglycerin in making suicide bombs in the early 2000s. Another nitrate ester, PETN, was blended with cyclonite to make a plastic explosive mixture used to destroy Pan Am Flight 103 in 1988 over Scotland. Residues of these same explosive ingredients were also identified in a terrorist event in India in 2003.

In 1867, Swedish engineer Alfred Nobel discovered that pure nitroglycerin could be absorbed into a porous material such as siliceous earth. The resulting mixture became known as **dynamite**. Nobel made a substantial fortune from his discovery and used it to establish a monetary fund known as Nobel prizes. Dynamite could be handled much more safely, a property that allowed it to be transported more easily and used with far less risk. Today, dynamite is manufactured by absorbing a mixture of nitroglycerin and diethylene glycol dinitrate into wood pulp, sawdust, flour, starch, or similar carbonaceous materials. Although itself an explosive, the purpose of diethylene glycol dinitrate is to depress the solution's freezing point, so that it remains a liquid at low temperatures. Calcium carbonate is added to neutralize the presence of nitric acid in case of hydrolysis.

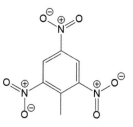

**FIGURE 14.2**   Molecular structure of TNT.

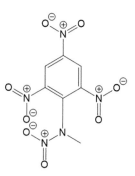

**FIGURE 14.3**   Molecular structure of tetryl.

Other explosives behave in much the same way and for much the same reasons. Consider trinitrotoluene, i.e., TNT, shown in Figure 14.2, and tetryl shown in Figure 14.3.

Trinitrotoluene or TNT has the chemical formula of $C_7H_5(NO_2)_3$ and is liquid at room temperature. It is prepared by reacting toluene with a mixture of nitric acid and sulfuric acid. Some of its properties are listed in Table 14.5.

The explosive nature or meta-stability of TNT can be seen in the stereochemistry of the three nitro groups bonded to the toluene molecule in Figure 14.2. Note that two molecules of *liquid* reactant decompose into 15 molecules of *gaseous* product plus seven atoms of solid carbon (as soot), similar in theme to the nitroglycerin reaction. The decomposition reaction of TNT is

**TABLE 14.5**
**Properties of TNT**

| | |
|---|---|
| Melting point | 178°F (81°C) |
| Boiling point | 563°F (295°C) |
| Detonation temperature | 450°F (232°C) |
| Specific gravity | 1.65 |
| Vapor pressure | 0.046 mmHg |
| Detonation velocity | 22,600 ft/s (6900 m/s) |

---

**TABLE 14.6**
**Properties of Tetryl**

| | |
|---|---|
| Melting point | 266°F (130°C) |
| Detonation temperature | 369°F (187°C) |
| Specific gravity | 1.60 |
| Detonation velocity | 23,300 ft/s (7100m/s) |

---

$$2C_7H_5(NO_2)_3(l) \rightarrow 7CO(g) + 5H_2O(g) + 3N_2(g) + 7C(s) \qquad (14.46)$$

It is relatively safe to handle. While it is a highly powerful explosive, it is not usually sensitive to heat, shock, or friction. It does not react with atmospheric moisture. Even after being kept in storage for several years, it is not susceptible to spontaneous decomposition.

Tetryl, or properly named 2, 4, 6-trinitroophenyl-N-methlynitramine, is a yellow solid. Some of its properties are listed in Table 14.6.

Note its similarity in molecular structure with nitro groups and potential meta-stability to nitroglycerin and TNT. Its detonation can be written as follows:

$$2(NO_2)_3 C_6H_2N(NO_2)CH_3(s) \rightarrow 11CO(g) + 5N_2(g) + 5H_2O(g) + 3C(s) \qquad (14.47)$$

Unlike nitroglycerin and TNT, tetryl is a solid at room temperature. The detonation of tetryl produces approximately the same explosive power as the same mass of TNT. When tetryl is blended with a mixture of liquid TNT and a small amount of graphite, the popular explosive called tetrytol is produced. Tetrytol is occasionally used by the military as the bursting charge in artillery ammunition. Like nitroglycerin, it is potentially toxic by inhalation, ingestion, or skin contact.

## 14.7   GREEN CHEMISTRY

Times Beach (1983 est. pop. 2000) is a small town in Missouri, about 25 miles southwest of St. Louis. Settled in the 1920s as a summer resort for working-class St. Louis residents, Times Beach gradually became a permanent community of small homes and trailer parks. In 1982, the soil along the roads of Times Beach was found to be contaminated by the toxic chemical dioxin. The most likely source of the contamination was the spraying of roads with dioxin-tainted waste oil about a decade earlier. Times Beach was one of 26 towns and 100 sites sprayed with such oil. The concentration of dioxin found in the soil varied anywhere from 300 to 740 parts per billion. The problem was aggravated when a flood in 1982 forced about 700 families to abandon their homes. Eventually, the federal government had to arrange the purchase of the entire town with a specially created "toxic waste clean-up" fund of $33 million and relocated the families involved.

Love Canal, Niagara Falls is located in upstate New York. In 1942, Hooker Chemical Company (which later became Occidental Chemical Corporation) began dumping its industrial wastes into an abandoned canal bed, Love Canal. The wastes consisted of chlorobenzenes and halogenated pesticides. Estimates of disposal range up to 21,000 tons. This practice continued until 1952 when it covered the site with a layer of topsoil. The city of Niagara Falls was given the land by Hooker Chemical in 1953 and proceeded to build a housing tract along with a new school. All seemed well until, in 1971, chemical substances began leaking through the clay cap that had sealed the dump years earlier. At least 82 chemicals were detected and catalogued, including a number of carcinogenic and teratogenic compounds, from benzene to chlorinated hydrocarbons to dioxins. Over the years, this area had experienced an unusually high number of liver cancer cases, birth defects, and miscarriages, as well as seizure-inducing nervous disease. It was finally declared an official disaster area. Over 1000 families had to be relocated. The state of New York paid $10 million to purchase some of the homes and another $10 million to contain the leakage.

These two tragic events underscore the failure of the management of chemical hazardous wastes. They also explain the concept and need behind "green chemistry" to become the new backbone of the chemical industry. Green chemistry seeks to minimize both the use of hazardous starting materials and the production of hazardous materials by seeking alternate reaction routes for synthesis of a desired product. The objective is to stop environmental pollution and exposure to hazardous materials by eliminating them at the source. A more formal definition of **Green Chemistry** is:

> Green chemistry is the use of chemistry techniques and methodologies that reduce or eliminate the use of feedstock, solvents, or reagents, or generation of products and by-products, that are hazardous to human health or to the environment.

This goal has several directives. One is to find processes that use the least amount of external energy, especially from the combustion of fossil fuels, as discussed in Chapter 8. One solution is finding a suitable catalyst that may lower the activation energy of a given reaction and thus replace the thermal heat required to run the reaction. Microwaves and sonic energy have also shown potential to minimize the need for thermal heat. In addition to running the reaction, purification and separation are also critical steps in the synthesis of a product and are similarly energy-intensive processes. These steps often require distillation, recrystallization, and ultrafiltration. By designing a process to minimize the need for these steps, and still achieve high degrees of purity of product, the chemist may be able to reduce the need for additional thermal or electric energy.

Another directive is the careful use of starting materials or feedstocks. Polymers are a very important class of compounds that have broad properties and applications. Starting materials for polymers, i.e., monomers, can frequently pose the full range of chemical hazards, when they are derived from petroleum or other fossil fuels. Polysaccharide feedstocks, on the other hand, have several advantages. They come from biological sources—biomass—which are renewable. To date, there are also no toxicological data to suggest that they pose significant acute or chronic toxicity to human health or to the environment. An added bonus is that they pose a lower risk in accident potential.

Another consequence involves toxicity considerations. Consider two starting reagents or feedstocks, X and Y. Both substances are *generally recognized as safe (GRAS)* except that each can cause blindness in humans. Substance X causes blindness when exposure exceeds an air concentration of more than 10 parts per billion, while substance Y causes blindness when exposure exceeds 1000 parts per billion. *Evaluation*: Substances X and Y have the same toxic effect, but substance Y is preferable because it is 100 times less potent in causing blindness. Hence, all other factors being equal, substance Y is the better choice from a green chemistry evaluation standpoint.

In the preceding example, the effect of only potency is considered. Similar evaluations can be made by comparing individual toxicity concentrations for any one of three, given, toxicity effects—potency, severity, or reversibility. The same argument can also be made for products, by-products, solvents, or even catalysts. In the end, a healthier workforce means a more productive workforce and a safer environment for all.

An actual case on point may better illustrate the message. Monsanto Company recently developed an alternative method, safer and more environmentally friendly or "green," to the synthesis and production of polyurethanes and their isocyanate precursors. The old method reacted an amine, $RNH_2$, with phosgene (carbonyl chloride), $COCl_2(g)$, an extremely, acutely toxic gas:

$$RNH_2 + COCl_2(g) \rightarrow RNCO + 2HCl \rightarrow RNHCO_2R \left( \text{extracted in R'OH} \right)$$

In their new synthesis, phosgene has been replaced by carbon dioxide, $CO_2(g)$:

$$RNH_2 + CO_2(g) \rightarrow RNCO + 2H_2O \rightarrow RNHCO_2R \left( \text{extracted in R'OH} \right)$$

There are two, often understated, benefits of green chemistry process. The first is efficiency, due to the elimination or minimization of undesired by-products, i.e., waste products, which must be removed and disposed. The second is potential reduction in reaction and net process times. Both of these should result in lower costs and higher profitability. Coupled with a healthier and safer working environment, this should translate into higher worker productivity and reduced health care expenses.

"Principles of Green Chemistry" can be expressed or summarized in different versions. One good version or source is *Green Chemistry: Theory and Practice* (Anastas and Warner 1998) whose principles are adapted and listed in Table 14.7.

## 14.8   NANOTECHNOLOGY

A frequently heard term involving modern technology is nanotechnology. The name comes from the size of the objects that are required for the technology to function as it does. Nanotechnology is built on the order of 1–100 nm in size. Sometimes many individual objects will be used together, making the overall size of the device much bigger, but it is the smallness of the building blocks that drives function.

Common examples of nanotechnology are substances such as graphene, carbon nanotubes, and quantum dots. Graphene is a single layer of graphite. It is a sheet of carbon that is a single atom thick. Graphene is tremendously strong and a conductive material, making it attractive for many potential applications. If one were to roll up a sheet of graphene and make a tube, one would get a carbon nanotube (although

**TABLE 14.7**

**Principles of Green Chemistry**

1. *Prevention*: It is better to prevent waste than to treat it or clean it up after creating it.
2. *Atom Economy*: Synthetic methods should be designed such that all materials used are maximized in the process of the final product.
3. *Less Hazardous Chemical Syntheses*: Wherever possible, synthetic methods should be designed to use and generate substances that possess little to no toxicity to human health and the environment.
4. *Designing Safer Chemicals*: Chemical products should be designed to affect their desired function while minimizing their toxicity.
5. *Safer Solvents and Auxiliaries*: Minimize the use of solvents and separation agents whenever possible.
6. *Design for Energy Efficiency*: Energy requirements should be recognized for their environmental and economic impact and should be minimized. Whenever possible, synthetic methods should conducted at room temperature and pressure.
7. *Use of Renewable Feedstocks*: A raw material or feedstock should be renewable instead of depleting natural resources whenever feasible.
8. *Reduce Derivatives*: Steps requiring the use of blocking groups, etc., should be minimized, since such steps require additional solvents and reagents, which generate waste. Waste often means lower product yield plus higher treatment/purification costs, which equals lower profitability.
9. *Catalysis*: Catalytic reagents are preferable to stoichiometric reagents.
10. *Design for Degradation*: Chemical products should be designed such that they break down or degrade into innocuous compounds and do not persist in the environment.
11. *Real Time Analysis for Pollution Prevention*: Analytical methods need to be developed for real-time, in-process monitoring and control of the formation of hazardous chemicals.
12. *Safer Chemistry for Accident Prevention*: Raw materials and feedstocks as well as the process to manufacture a desired product should be chosen to minimize the potential of chemical accidents, including release of toxic products, fires, and explosions.

this is not how they are usually manufactured). Carbon nanotubes also have very interesting strength and conduction characteristics. If woven into a cable, it is thought that carbon nanotubes could be strong enough to allow for the construction of space elevators. Quantum dots are metallic particles grown to be a few nanometers across. These particles can be used to generate light and are promising for the development of newer display device technologies with improved color ranges compared to current technology.

Nanotechnology also has potential applications in energy storage. Capacitors are a form of energy storage that can be charged and discharged quickly, but tend to store less energy than batteries by weight. Batteries hold more energy, but cannot release it as quickly. Capacitors are constructed from two thin charge-carrying plates separated from each other by an insulator. A limitation of this technology is how small these plates can be constructed. With nanotechnology, significant improvements can be made in the sizes possible, improving the storage capacity of the capacitor. The hope is to be able to develop supercapacitors, which can hold as much energy as a battery, but discharge it much more quickly.

Another good example is nanoscale silver particles, which have a particularly strong antimicrobial effect. Nanometer-size silver mesh can be used as a filter, which will kill bacteria in water filtered through it. One class of masks marketed as "boomer" masks used in COVID virus protection contains nanoscale silver particles infused into the cloth. It is washable and reusable.

Many of these materials are poorly understood from the perspective of safety, and the new properties arising from their smallness are sometimes a cause for concern. For instance, carbon nanotubes have been noted to display properties similar to asbestos when they are in their loose form, potentially causing cancer in people inhaling particles of carbon nanotubes. However, many other types of nanoparticles have unknown environmental impacts. Many forms of nanotechnology have their components tightly bound in place and are unlikely to be a health risk. Loose particles, however, have the potential to be airborne and could be easily absorbed by plant or animal life. Their impact remains in question. Research and regulation in this field face tough challenges.

## 14.9   THREE IMPORTANT "ENVIRONMENTAL" DISASTERS

### 14.9.1   West Virginia MCHM Leak of 2014

On January 9, 2014 at 6 PM, Governor Earl Ray Tomblin of West Virginia gave some 300,000 residents of Charleston unexpected, bad news. Their drinking water had just been contaminated with an obscure chemical used in the coal processing industry by the name of MCHM—"4-methylcyclohexanemethanol." His immediate order was to avoid drinking any tap water. The CDC (Center for Disease Control, an arm of the US National Institutes of Health) emphasized this ban for pregnant women 6 days later, on January 15. This ban would later be extended to include bathing and washing clothes. The only sign of contamination reported by many residents was a telltale smell of licorice coming from their laundry rooms. Residents were advised to drink only bottled water until further notice (Figure 14.4).

The source of the chemical was a leak from a storage tank, tank number 396 of Freedom Industries, located near the Elk River, about 1 mile upstream from the intake pipe of the Charleston water supply. Initial sampling on the Elk River showed 1.04–3.35 ppm MCHM near the intake pipe but dropped to 1.02–1.56 ppm in the treated drinking water [C&EN; 2/17/14; p 10]. Tragically, there was only one 2011 MSDS available from its manufacturer, The Eastman Chemical Company, which warned only that the "substance may be harmful if swallowed" and causes "skin and eye irritation." It offered little to no information regarding inhalation hazards, its reproductive or chronic toxicity, carcinogenicity, biodegradability, or its general physical and chemical properties. Acute toxicity data included two entries: oral $LD_{50}$ (rat) = 825 mg/kg and dermal $LD_{50}$ (rat) >2000 mg/kg. In other words, the Eastman Chemical Company MSDS showed more than two dozen entries marked as "no data available."

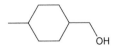

**FIGURE 14.4**   4-Methylcyclohexanemethanol or MCHM.

CDC used the results of a 28-day study of pure MCHM on rats, which showed that a daily dose of 100 mg of pure MCHM per kg of body weight produced no observable effects in rats. Assuming a child of mass 10 kg drinking 1 liter of water per day, and allowing for a safety factor of 10 (100 or 1000 is more common), CDC concluded that a concentration of 1.0 ppm or less in drinking water would show no observable adverse effects and would be safe for the residents.

On January 17, 2014, Eastman Chemical voluntarily released toxicology studies it had conducted in the 1980s and 1990s on crude and pure MCHM. The studies included both human and environmental exposure, though on a random, limited scale.

On January 21, Freedom Industries reported that a second chemical, propylene glycol phenyl ether, PPH, was also contained in the same tank, as part of a proprietary mixture, at a concentration of 7.3% by mass. By most accounts, it was less toxic than MCHM. Initial estimates of the spill on January 11 were 7500 gals, but this figure was later revised to upward of 10,000 gals of a crude MCHM and PPh blend. The storage tank capacity was 25,000 gals.

CDC testing of water samples taken from both the Elk River and treated drinking water continued through the month of January and started to show signs of dropping MHCM concentrations. By early February, concentrations of MCHM continued to abate with only a trace amount showing up in drinking water, e.g., 0.013–0.018 ppm in samples taken from elementary school drinking water fountains. CDC gave the *all clear* to drink the tap water for everyone, including pregnant women, on February 5.

It should be noted here that the then 38-year-old federal law—The Toxic Substances Control Act of 1976—governing the production and environmental release of such chemicals, did not expressly require chemical manufacturers to provide this information. They were required to submit hazard information to the EPA *only* "when data suggest the possibility of substantial risk." Instead, TSCA set up complex legal requirements that EPA must meet before it can require manufacturers to provide toxicity data for chemicals in commerce. In addition, MCHM was actually in production in 1976 at the time when TSCA was enacted. But like 62,000 or so other commercial chemicals already in production at that time, lawmakers qualified it to be grandfathered in, avoiding the testing and reporting requirements of TSCA. Further, since MCHM was also not among the 2200–2800 HPV (high production volume) chemicals at the time, it did not fall under the scrutiny of the EPA 1998 "voluntary program agreement" with the American Chemistry Council and the Environmental Defense Fund. However, it later became a HPV chemical in 2002, and at the time of this accident, Eastman acknowledged it was producing between 5 and 10 million pounds per year of crude MCHM.

An October 2015 report based on updated studies was released by the National Toxicology Program (NTP). The NTP is an interagency program established in 1978 to coordinate toxicology research and testing across the US Department of Health and Human Services. It is supported by three core agencies: the National Institutes of Health (NIH) and the National Institute of Environmental Health Sciences (NIEHS), the US Food and Drug Administration (FDA), and the NIOSH and the CDC. The report stated that the spill into the Elk River in January of 2014 consisted of 89% crude MCHM, 11% propylene glycol phenyl ethers, and 4% water. The product called crude MCHM is a commercial mixture containing >70% MCHM along with lesser concentrations of five other chemicals (MMCHM, MMCHC, DMCHDH, CHDM, and 2MCHM).[2] The propylene glycol phenyl ether component is also a commercial mixture containing predominantly DiPPH and PPH in unknown concentrations. Laboratory studies of relatively short duration using rodents and other model organisms, as well as computer modeling, were conducted throughout 2014 and 2015. These studies were designed to evaluate the toxicological effects and identify what biological systems (e.g., liver, kidney, blood) were affected, for both short- and long-term effects. Studies were carried out on both major and minor constituents of the chemical spill, using computer modeling where necessary and appropriate. Preliminary results from these studies are available but need to be peer-reviewed. Research results may be accessed by contacting the NTP.

### 14.9.2   BRITISH PETROLEUM AND THE *DEEPWATER HORIZON* ACCIDENT OF 2010

The *Deepwater Horizon* accident in 2010 is a major recent oil spill. In this accident, a deep ocean wellhead ruptured, and crude oil escaped from the wellhead for 87 days before it was closed. It is estimated that 210 million gallons of crude oil entered the Gulf of Mexico in this time. Oil spills of this sort have massive environmental impact. As oil is less dense than water, the oil tends to rise to the surface, and wildlife that lives on or near the surface can become covered in crude oil. This exposure proves to be a mechanical hindrance to many animals, for instance, preventing birds from flying, and reduces the insulation abilities of both fur and feather. Moreover, crude oil tends to have many toxic compounds in it, which kill fish, birds, and other aquatic life.

The cleanup of an oil spill is both a mechanical and chemical challenge. Although oil spills tend to come to the surface due to their lower density, they also tend to make thin layers not easily isolated from water. Furthermore, although oil is not generally soluble in water, small amounts of some of the toxic components of crude oil do dissolve into the surrounding water, making the water itself toxic as well.

One clever technology being developed to fight such spills is special sponges that selectively absorb oily compounds but not water. A number of such products have recently been created in response to the disaster. These materials tend to be highly porous materials that have strong interactions with oily compounds, but are chemically engineered to be particularly hydrophobic. In addition, their sponginess means that they can be easily reused, as they need merely to be squeezed out, and are ready to absorb more oil.

The majority of the *Deepwater Horizon* cleanup effort was actually done with surfactants, one by the brand name of Corexit in particular. Corexit, and surfactants like it, are composed of long chains of hydrocarbons attached to a central functional group that has several strongly polar bonds on it (see Section 1.4). The long chains

---

[2] The chemical abstract services registry number (CASRN) for each chemical substance listed above is included in parentheses after its acronym. The CASRN is an identification number unique to every known chemical substance. It lists the chemical and physical properties of a substance. No two chemical substances may have the same CASRN.

**FIGURE 14.5** Surfactant action: As shown in this image, surfactants like Corexit break up oil to make it water-soluble, but all the oil is still present, just in smaller, dispersed form.

can easily pass into solution within an oil, while the polar bonded area of the molecule prefers to be dissolved in the water layer. This leads to the interesting case of one molecule being dissolved at the interface of water and oil layers, the hydrophilic part of the surfactant in the water, and the hydrophobic part of it in the oil. The tendency for these molecules as a group is to then form a micelle, a spherical construct with the hydrophobic chains in the middle, and the hydrophilic groups covering the surface. These micelles can then trap and carry away droplets of oil in the middle, as the smaller droplet of oil is now covered in water-soluble hydrophilic parts of the surfactant (Figure 14.5).

As shown in this image, surfactants like Corexit break up oil to make it water-soluble, but all the oil is still present, just in smaller, dispersed form.

The use of Corexit, of course, leads to the disappearance of surface patches of oil, as the spills become solubilized in water. The question of what happens to the oil is important, as the oil is now in much smaller particles but dispersed in the ocean. Significant argument can be had over whether this approach is a suitable way to deal with spills, but it is significantly more cost-effective than trying to recover all of the spilled oil. Studies continue to show, however, that the dispersed oil is still affecting the health of aquatic life in the Gulf, with potential impact on the local fishing industries, and even the health of people consuming seafood from the Gulf.

### 14.9.3 UNION CARBIDE AND BHOPAL, INDIA OF 1984—THE WORST CHEMICAL ACCIDENT EVER

Methyl isocyanate (MIC) was a chemical used to make a pesticide known as carbaryl (a carbamate) in a one-step process by reacting it with 1-naphthol—see Figures 14.6 and 14.7. MIC is weakly soluble in water (10 parts MIC per 100 parts water) but will react with water exothermically to produce carbon dioxide gas.

**FIGURE 14.6** Molecular structure of methyl isocyanate or MIC.

**FIGURE 14.7**  Molecular structure of 1-napthol.

On December 2, 1984, MIC was kept in special storage tanks to prevent it from coming into contact with water. The storage tanks were owned and operated by Union Carbide, a chemical company headquartered in the Danbury, CT (USA), but operating a manufacturing plant in Bhopal, India. Union Carbide was one of 30 companies then listed on the Dow Jones 30 industrial index. On that day, workers were cleaning out the pipes near the MIC tanks. It is unclear whether one of the valves to the tank failed and started to leak, or one of the workers neglected to keep the necessary valves properly shut during service, but water started to enter one of the MIC tanks. To make a bad situation worse, the tank, which by company safety regulation is to be kept only 50% full, was over 75% full. The purpose of the unfilled space was to provide a buffer in case of an accident. Water slowly leaked in and a chemical reaction ensued, causing the temperature and pressure in the tank to increase. At first, workers realized that MIC was leaking when they felt their eyes burning, but were not sure that something major was wrong. This condition was reported to the supervisor who subsequently checked the pressure and found it to be within the normal range at 10 psi. However, when the pressure was checked about 1 hour later, it had risen to 55 psi (well above the safe range) and the temperature was 25°C. The refrigeration system was down because the coolant had been recently drained. In less than another hour, the pressure shot up to 180 psi and the temperature increased to 200°C. The tank ruptured and a plume of gas shot up into the air lasting nearly 2 hours. The plume was mostly MIC—36–40 metric tons. By this time, it was almost midnight and into the next day.

It is important to note the problem was not that the chemical reaction had produced a toxic gas. Rather, the chemical reaction between MIC and water had created conditions of high temperature and pressure, which made the storage tank rupture, releasing the unreacted MIC as a toxic gas into the atmosphere. MIC has a low boiling point (39.5°C) and is very volatile, so virtually all of the liquid MIC was converted into a gas by the time of the explosion. MIC reacts with water to produce methyl amine according to the following reaction, releasing a large amount of heat, as indicated by the large, negative ΔH value:

$$CH_3 NCO(g) + H_2O(l) \rightarrow CH_3 NH_2(g) + CO_2(g) \qquad \Delta H = -566 \, kJ/mol$$

MIC vapor is highly toxic by inhalation and skin adsorption, irritating to the nose and throat, and corrosive to skin. It causes dizziness, coughing, shortness of breath, and most importantly, pulmonary edema, leading ultimately to death. As result, more than 200,000 people were immediately exposed to the MIC vapor, and perhaps as many as 500,000 over the long run. It was estimated that between 2500 and 3800 people died from exposure within a few days. Tens of thousands suffered from lung cancer and kidney failure and later died.

The chief causes of the accident were identified as follows:

- Storing large amounts of MIC in tanks beyond recommended levels
- Insufficient and infrequent maintenance, which explains the failure of safety systems
- Switching safety systems off to save money, e.g., shutting down the refrigeration system on MIC storage tanks
- Possible sabotage of a few disgruntled employees; outright negligence on the part of others
- No viable emergency response plan, team, training, or equipment in place in the event of an accident
- Locating a large number of shacks housing indigent residents too close to the plant

The Indian government had stipulated that the plant be operated exclusively by Indian workers. Thus, Union Carbide had agreed to train Indian workers and thus flew in Indian workers to a sister site in West Virginia for hands-on training sessions, but practiced an otherwise hands-off attitude toward the Indian managers. The company required that US trained engineering teams make both periodic and occasional, unannounced on-site inspections for safety and quality control. However, these inspections were stopped in 1982, when the plant decided it was too expensive and unnecessary. In fact, the last US inspection in 1982 warned of many hazards, which routine safety inspections and maintenance would correct. From 1982 until 1984, safety measures and quality control efforts declined, presumably due to a large turnover of employees, inadequate and improper training of new employees, and a general lack of technical savvy among the local workforce. Even though many of these employees would complain about headaches and nausea, they refused to wear proper PPE such as gloves and respirators, because they said that it was too hot and uncomfortable. Indeed, Bhopal has a subtropical climate.

Though American lawyers flew to India and were eager to help families of the victims of the Bhopal disaster, US District Judge John Keenan of New York dismissed the multibillion lawsuit against Union Carbide, citing that more appropriate legal action could be taken in the Indian court system, because that was where the bulk of the evidence and victims could be located.

Litigation shifted to India. In response to $3 billion asked for by the Indian government, Union Carbide offered only $200 million. In the end, the Indian government settled for $470 million (equivalent to $900 million in today's dollars). That was the end of the story for Union Carbide. Dow Chemical bought the company in 1999, denying any liability for the 1984 disaster during its acquisition. But then, in 2010, 26 years after the disaster, eight mid-level Indian managers who had worked for Union Carbide were convicted in an Indian court of criminal negligence and were sentenced to short prison terms.

MCHM (34885-03-5): 4-Methylcyclohexanemethanol
2MCHM (2105–40-0): 2-Methylcyclohexanemethanol

MMCHM (98955-27-2): 4-(Methoxymethyl)cyclohexanemethanol
MMCHC (51181-40-9): Methyl 4-methylcyclohexanecarboxylate
DMCHDH (94-60-0): Dimethyl 1,4-cyclohexanedicarboxylate
CHDM (105–08-8): 1,4-Cyclohexanedimethanol
DiPPH (51730-94-0): Dipropylene glycol phenyl ether
PPH (770-35-4): Propylene glycol phenyl ether

# 15 Introduction to Basic Toxicology

## 15.1 THE PROBLEM

In 2015, Dr. David Perlmutter wrote *Brainmaker* (Little Brown. Boston. 2015) in which he summarized the following findings and facts:

- More than 100.000 synthetic chemicals in the marketplace approved for use in the United States.
- About 84,000 industrial chemicals in use each year in the United States.
- Approximately 1000 different chemicals used in pesticide formulations.
- Over 9000 chemicals used as food and beverage additives either have flavor enhancers or shelf-life extenders.
- Approximately 3000 chemicals used in the manufacturing of pharmaceuticals.
- Roughly 3000 chemicals used in the manufacturing of cosmetic products.
- Only 200 chemicals out of 84,000 have actually been tested for toxicity under the Toxic Substances Control Act (TSCA) or regulated by FDA.
- In total, 800 chemicals are known to be lipophilic and interfere with the human endocrine system.
- In total, 8000 chemicals are produced at more than 25,000 lbs. per year in the United States.

In an article published in the February 17, 2020 issue of Chemical and Engineering News (citing a report in *Environ. Sci. Tech.* 2020, http://pubs.acs.org/DOI:10.1021/acs.est.9b06379), the following global inventory of chemicals, thus far the most comprehensive, is reported:

- More than 350,000 chemicals or mixtures of chemicals are in use today, which breaks down into:
- 157,000 individual chemicals identifiable by CAS numbers (CASRN).
- 120,000 substances that could not be conclusively identified.
- 75.000 mixtures, polymers, and substances of unknown or variable composition.

In a recent article published in Science (Tian et al. 2020), researchers at the University of Washington report that they have pinpointed the single, chemical substance that leaches out of from rubber tires and then contaminates rivers and streams in the Pacific Northwest accounting for massive die-offs of coho salmon before they have a chance to spawn. Roadway runoff includes tiny rubber particles, which shred off from rubber–road friction. Using HPLC/MS as a diagnostic tool, the single chemical identified is 6 PPD-quinone, a compound used to retard the degradation of tires. Interestingly, this compound was not found in the database of environmental chemicals or in the published characterizations or compositions of rubber. This underscores the magnitude of the problem.

More conservative estimates by American Chemistry Council (the trade organization of the US chemical industry) say that the US chemical industry produces around 100,000 different chemicals annually. Of these, 2800 chemicals are considered high production volume (HVP) chemicals and are closely watched by the EPA. HVP chemicals are produced at the rate of at least 1 million pounds per year. This list *does not* include polymers, e.g., plastics or metals.

In 2012, The Natural Resources Defense Council (NRDC) noted that "more than 80,000 chemicals permitted and in use today in the United States have never been fully assessed for toxic impacts on human health and the environment." When TSCA and Resources Conservation and Recovery Act (RCRA) were initially passed in tandem in 1976 (though TSCA has been recently upgraded in 2016 as The Frank Lautenberg Law mentioned in Chapter 14), the intention was to safeguard and protect public health through monitoring the life cycle of all chemicals manufactured and used in products sold and consumed in the United States. At that time, there were about 62,000 chemicals that were allowed to remain on the market without testing for their health effects or impact on the environment. Up to 2016 (when a long overdue overhaul of the 1976 TSCA was signed into law in June of 2016, known formally as the Frank R. Lautenberg Chemical Safety for 21st Century Act, finally giving the EPA authority to request safety data on chemical substances that are both new and already in use and in the marketplace), in the almost 40 years that have transpired under these two statues, the EPA has required testing of only about 200 of those chemicals and has succeeded in regulating only five under TSCA and RCRA. While it is true that other laws have set standards for clean air, clean water, safe drinking water, and workplace safety, they do not address or regulate exposure to most chemicals broadly encountered by American consumers and workers.

Fortunately, under the 2016 updated TSCA law, ten new chemicals were identified and designated to be assessed and evaluated by June 2020, particularly as they posed hazards to workers' health (traditionally covered by OSHA regulations). Those chemicals are 1-bromopropane; carbon tetrachloride (a common feedstock); hexabromocyclododecane; methylene chloride; asbestos; 1,4 dioxane; N-methylpyrrolidone; tetrachloroethylene; trichloroethylene; and Pigment Violet 29. Seven of these chemicals are chlorinated organic solvents. That deadline unfortunately was missed. As of November 2020, only the first four have been formally assessed. Assessments for the remaining six are set to be completed by the end of 2020. A longer list of top 20, which includes four flame retardants, five phthalates, 1,3 butadiene, a fragrance additive, and formaldehyde, is also yet to be completed.

The objective of these two laws was to assess the hazards of commonly encountered chemicals, particularly those manufactured on a grand scale, from cradle to grave. Further, since 1976, chemical manufacturers have introduced at least 22,000 new chemicals into commerce, for which they have provided little to no information regarding their potential effects on human health or impact on the environment. Many of the chemicals in question are found in toys and products designed for use by children, cleaning and personal care products, pesticide formulations, electronics, building materials, food and beverage additives, fabrics, furniture, and automobile interiors. Two common chemical categories reflected in these products are plasticizers and flame retardants. It should be stressed that once a synthetic chemical is introduced into the marketplace, it is only a matter of time before it or one of its metabolites will find its way into some medium of the environment. This is called "environmental fate" and its *cumulative* consequence is called "**environmental burden**." For example, n-hexane, $C_6H_{14}$, a component of gasoline and many adhesives, is first metabolized by the liver to 2,5 hexanediol and then to 2,5 hexanedione, a potent neurotoxin. Eventually, some proportion of these molecules will be eliminated by the kidneys into the urine, which in turn will end in a municipal wastewater treatment plant before being released into the environment.

In a June 23, 2015 news release from Brunel University in England, a research team composed of 174 scientists from 28 countries examined 85 common chemicals believed to have no link with cancer. The investigators concluded, however, that at current exposure concentrations, *mixtures* of 50 of those same chemicals *can* lead to cancer. This is known as the "synergistic effect." Spot studies have shown that the synergistic effect is not linear.

"This research backs up the idea that chemicals not considered harmful by themselves are combining and accumulating in our bodies to trigger cancer and might lie behind the global cancer epidemic we are witnessing," said study coauthor Hemad Yasaei. He added: "We urgently need to focus more resources to research the effect of low-dose exposure to mixtures of [synthetic] chemicals in the food we eat, air we breathe, and water we drink." ("Getting to Know Cancer;" Global Taskforce assembled and led by Dr. Leroy Lowe of Halifax Nova Scotia.)

In fact, two areas of research on the toxicity of chemicals that have been notably neglected are as follows:

1. Laboratory studies of the net effect of exposures to a *combination* of chemicals taken simultaneously in terms of dose–response.
2. Laboratory studies of the difference between male and female dose–response relationships to a given toxin or toxicant; preliminary studies indicate there *is* a significant difference.

For readers who might think that the net effect of exposure to these toxicants has negligible consequences or impact on overall public health and associated healthcare costs or costs due to lost wages, please consider the summary findings of the November 2017 Lancet Commission Report. Below are some important highlights:

- Pollution linked to 9 million deaths worldwide in 2015.
- *Air pollution* shown in a global map (and sources vary greatly) is the biggest contributor, linked to 6.5 million deaths, while water pollution is the second contributing to 1.8 million deaths.

- *Workplace-related pollution* linked to 800,000 deaths (violations of labor standards).
- Almost all (92%) occur in low- and middle-income countries and in rapidly industrializing countries, e.g., India, China, Pakistan, Kenya, and Madagascar.
- **According to the** World Health Organization **(WHO), Welfare (Health Care Costs) losses due to pollution estimated to cost more than $4.6 trillion each year, equivalent to 6.2% of global economic output.**
- Most of the 9 million deaths are due to noncommunicable diseases—heart disease, stroke, lung cancer, COPD.
- Endocrine-disrupting chemicals (EDC, see Section 15.8) contribute to disease and dysfunction and incur high associated costs, meaning >1% of the GDP of a given country or region.
- EDCs include solvents and lubricants and their by-products such as polychlorinated biphenyls (PCBs) (banned in the United States in 1979), polybrominated diphenyl ethers (PBDEs), dioxins, bisphenol A (BPA), phthalates (DOP, DEHP), discussed in more detail in Section 15.8.
- And pesticides such as atrazine, chlorpyrifos, methoxychlor, lindane, and DDT (banned for use in the United States in 1972, but still shows up in crops, soil, aquatic life, even in breast milk today).
- Diseases and adverse health effects associated with EDCs include prostate and breast cancer, infertility, male and female reproductive dysfunction, birth defects, diabetes, obesity, neurobehavioral and learning dysfunctions (such as ADHD and ASD), and autoimmune disorders.
- As countries industrialize, the type of pollution changes. Deaths from ambient air pollution, soil pollution, and occupational exposures have increased from 4.3 million (9.2%) in 1990 to 5.5 million (10.2%) in 2015. **STUNNING RESULT! Does this mean that economic development and progress must come at the expense of public health?**

### 15.1.1  FIVE IMMEDIATE CONCLUSIONS

- The disease costs of EDCs higher in the United States than in Europe:
- $340 billion or 2.33% of GDP in the United States compared with $217 billion or 1.28% of GDP in European Union, just under a ratio of 2:1.
- Difference driven by loss of IQ points and intellectual disability due primarily to PBDEs in the United States.
- In total, 11 million IQ points lost and 43,000 cases costing $266 billion in the United States vs. 873,000 IQ points lost and 3,290 cases, costing $12.6 billion in European Union.
- The single largest contribution to the loss in Europe came from organophosphate pesticides at $121 billion, while the US costs in this area were only $42 billion.

While keeping in mind that:

- US Pop: 330 M ⇔ EU Pop: 508 M ⇔ Europe Pop: 741 M.

## 15.2 BASIC CONCEPTS

Toxicology is the study of poisons and the harmful effects of chemicals on animals including humans and the environment. These chemicals may be naturally occurring or synthetic. If a chemical is harmful, it is said to be toxic and referred to as a **toxicant**. This should be differentiated from the term **contaminant**. Contaminant refers to a chemical that is not ordinarily present in the substance, mixture, or matrix of concern. It may or may not be toxic.

If a toxicant is a naturally occurring substance, it may more specifically be referred to as a **toxin**. **Poison** is a more common term for toxicant and refers to any substance that is toxic, whether natural or synthetic, regardless of any of the three, main exposure routes resulting in absorption: **inhalation, ingestion,** or **dermal (or eye) contact**. Dose (in bold print) is the amount of chemical entering the body. Absorption means the ability of the toxicant to enter the bloodstream at some point, though it may be first be absorbed into the lungs (inhalation), skin (dermal), or GI tract (ingestion). It is presumed that the blood is in equilibrium with the body's tissue and cells. If a substance is toxic by involuntary injection, it is referred to as a venom. Some snake and spider bites are venomous. Intravenous (or IV) is a fourth, less common absorption route.

One of the most important concepts in toxicology can be stated by the relationship:

$$Toxicity = Potency \times Exposure$$

This can also be expressed as follows:

$$Risk = Hazard \times Exposure$$

A toxic effect is an adverse response in an organism caused by exposure to a chemical. Potency generally describes the rate at which a chemical causes effects. A more potent chemical has a higher rate than a less potent chemical. For example, one drop of strychnine, a natural chemical extracted from the seeds of *Nux vomica*, attacks the nervous system, i.e., is a neurotoxin, and may be lethal. It is a very potent chemical! But a teaspoon of sodium chloride, i.e., common table salt, dissolved in a glass of water could be quite easily tolerated. The difference in dose levels is a measure of potency. The potency of chemicals clearly varies greatly, as shown in Tables 15.1A and 15.1B. While the potency is never zero, the exposure to a chemical and its risk

## TABLE 15.1A
## Standard Categories of Toxic Substances

| Category | Concentration | Example Chemical |
| --- | --- | --- |
| Extremely toxic | $LD_{50} < 1.0$ mg/kg | Botulinum; dioxin (TCDD); ricin |
| Highly toxic | $LD_{50} = 1–50$ mg/kg | Aflatoxin; strychnine; hydrogen cyanide |
| Moderately toxic | $LD_{50} = 50–500$ mg/kg | DDT; acetaminophen; heroin |
| Slightly toxic | $LD_{50} = 500–5000$ mg/kg | Ethanol; malathion; aspirin |
| Practically nontoxic | $LD_{50} = 5000–15,000$ mg/kg | Alar |

**TABLE 15.1B**
**Select Toxic Substances and Their Concentrations**

| Chemical (In Humans) | $LD_{50}$ (mg/kg) | Chemical (in Animals) | $LD_{50}$ (with route and animal) |
|---|---|---|---|
| Ethyl alcohol | 10,000 | Caffeine | 620 mg/kg—oral mouse |
| | | | 192 mg/kg—oral rat |
| | | | 105 mg/kg—iv rat |
| | | | 68 mg/kg—iv mouse |
| Sodium chloride | 4,000 | | |
| Hydrogen cyanide | 2.0 | Chlorine gas (LC 50) | 293 ppm/1h—rat |
| | | | 137 ppm/1h—mouse |
| Methanol | 810 | THC (from marijuana) | 175 mg/kg—iv mouse |
| | | | 100 mg/kg—iv dog |
| | | | 1270 mg/kg—oral rat |
| Strychnine | 30-100 | | |
| Nicotine | 1.0 | Mercury (I) chloride | 210 mg/kg—oral rat |
| | | | 8 mg/kg—iv mouse |
| Pentaborane | 50 | Arsenic acid (V oxidation state) | 48 mg/kg—oral rat |
| Rattle snake | 0.24 | Arsenic trioxide (III oxidation state) | 20 mg/kg—oral rat |
| Dioxin (TCDD) | 0.001 | Dioxin (TCDD) | 3.0 mg/kg—oral hamster |
| | | | 0.0016 mg/kg--oral guinea pigs |
| Botulinum toxin (Botox) | 1.0 ng/kg (Estimated) | | |
| Isopropyl alcohol (2-propanol) | 1000 | Strychnine | 2.35 mg/kg—oral rat |
| | | | 0.582 mg/kg—iv rat |
| | | | 0.50 mg/kg--oral dog |
| | | | 0.80 mg/kg—iv dog |

*Sources:* Wikipedia: https://en.wikipedia.org/wiki/Median_lethal_dose; U.S. ATSDR: https://www.atsdr.cdc.gov/toxprofiles/.pdf.

can be controlled. Thus, if the exposure to a chemical is zero, there are no toxic effects because the product of potency times exposure is zero.

The route or site of exposure is important since it determines the absorption rate and effectiveness of the toxicant. In general, the effectiveness can be stated as follows:

IV > Inhalation > Ingestion > Topical

"Dose makes the poison" is a common phrase. Clearly, the amount of toxicant inhaled, ingested, or dermally contacted—the three main routes of exposure—is vital to know. Table 15.1A, adapted from NIOSH guidelines, provides a list of some common chemicals and identifies approximate amounts (doses) in terms of concentration ranges in a living body and their degree of toxicity arranged by five, standard categories. Table 15.1B provides a short list of additional chemicals and their toxicities as well as their routes of exposure in animals.

Here, the $LD_{50}$ value is the lethal dose by oral ingestion or dermal contact for 50% of the population of a given species. $LD_{50}$ values are usually expressed in terms of mg of toxicant per kg of body weight. The toxicant is administered as a one-time dose. This is a measure of acute toxicity; bioaccumulation is not accounted for in this table. It is important to note that the toxicity level for a combination of two or more chemicals is not listed because toxicological studies have not been conducted to establish these limits, except in rare circumstances, despite their obvious everyday relevance.

The sources of toxic chemical substances vary. Some of these chemicals are synthetic or manufactured, such as those used in adhesives and pesticides (e.g., formaldehyde, Alar, DDT, atrazine), while others are naturally occurring in terrestrial species, such as in reptiles and amphibians (e.g., cobra neurotoxin, frog kokol venom), in plants (e.g., ricin, nicotine, strychnine), in insects and microorganisms (e.g., tetanus toxin, aflatoxin), or in marine organisms (e.g., equinatoxin from the sea anemone). In many instances, confirmation of a toxin has come from bitter experience.

The questions that are most frequently asked when assessing risks with exposure to toxic substances are as follows:

- What specific hazards are associated with or related to exposure to the substance in question? Does it cause liver cancer, central nervous system (CNS) malfunction, breathing difficulty, or some other problem?
- What exposure concentration would an average person (or group of similar persons) see during an average lifetime? In order to answer this, all three exposure routes must be considered.
- What is the relationship between the dose received and the response observed? How much can the average person tolerate before the onset of symptoms? Are the toxic effects reversible? When is the onset of cancer first observed? These are difficult questions to answer because quantitative testing on humans would have to take place. This is not a desirable way to evaluate the effects of chemical substances, so the laboratory animal is introduced and studied. But how accurate the dose–response curve is, particularly at the low end, for cancer and other illnesses is not well documented. And how accurate the results are when the curve is extrapolated or transposed to human cases is also not well documented.
- Finally, what constitutes an acceptable risk? The usual answer is that the highest dose that produces no effect, termed the "no observable adverse effect level" (**NOAEL**) is the ideal limit of acceptable risk. An alternate answer is the standard that produces negligible risk, i.e., one case in 1 million people.

For comparison, mortality risk tables for some common, voluntary, and involuntary activities are given below. Figures are only *approximate* lifetime odds and are based on data available for persons born in the United States in 2001 and are offered for the sake of comparison only (Table 15.2).

It is often difficult to establish a link between chemical exposure in the environment and the onset of cancer because of the latency period, the lack of specific information on the concentration of exposure, i.e., the dose, and the frequent absence of a

**TABLE 15.2**
**Mortality Rates**

| Cause of Death | Lifetime Odds |
|---|---|
| Heart disease | 1 in 5 |
| Cancer (all forms) | 1 in 7 |
| Stroke | 1 in 23 |
| Accidental injury | 1 in 36 |
| Motor vehicle accident | 1 in 100 |
| Intentional self-harm (suicide) | 1 in 121 |
| Assault by firearm | 1 in 246 |
| Fire or smoke | 1 in 1116 |
| Drowning | 1 in 8942 |
| Air travel accident | 1 in 20,000 |
| Flood | 1 in 30,000 |
| Lightning strike | 1 in 83930 |
| Venomous bite or sting (snake or bee) | 1 in 100,000 |
| Earthquake | 1 in 131890 |
| Tsunami | 1 in 500,000 |
| Fireworks discharge | 1 in 615,488 |

*Sources:* National Center for Health Statistics, CDC; American Cancer Society; National Safety Council; International Federation of Red Cross and Red Crescent Societies; World Health Organization (WHO); USGS; NASA.

pure control factor. As mentioned earlier, laboratory animal testing is the best alternative. While highly controversial, the primary objective of toxicity testing in laboratory animals is, of course, to minimize potential harm from chemicals to humans. The specific objectives of laboratory animal testing include the following:

- Identifying the target organ(s) or systems affected by the chemical.
- Identifying which of the three routes of exposure is the most sensitive.
- Establishing if the effects are permanent or reversible.
- Determining the mechanism of toxic action.
- Determining the most sensitive method for detecting the toxic effect.
- Evaluating the $LD_{50}$ value (lethal dose by oral ingestion for 50% of a population of a given species) or $LC_{50}$ value (lethal concentration by inhalation for 50% of a population of a given species). $LD_{50}$ values are usually expressed in terms of mg of toxicant per kg of body weight. The toxin is administered as a one-time dose. This is a measure of acute toxicity. $LC_{50}$ value can refer to a toxicant dispersed in air or dissolved in water and inhaled or ingested, respectively.

Some chemicals are known to cause cancer in humans at specific, main tumor sites referred to as target organs. At the time of this writing, lung cancer is the third leading

**TABLE 15.3**
**Carcinogenic Chemicals**

| Chemical Name | Source | Main Tumor Site(s) |
|---|---|---|
| Arsenic | Natural rock; pesticides | Lung; skin |
| Benzene | Industrial solvent | Blood; bone marrow |
| DES | Fertility drug | Breast |
| Direct blue | Azo dye | Liver |
| Hexavalent chromium | Natural rock; paint pigment | Lung |
| Vinyl chloride | Synthesis of PVC | Lung; liver |

cause of cancer, behind prostrate and breast cancers, respectively. See Table 15.3 for more examples. Check Appendix V for a more detailed anatomical description of toxicants and their target organs.

## 15.3 MORE TERMS AND DEFINITIONS

Using generally accepted protocols, four different lengths or durations of time have been developed:

- **Acute tests**: a one-time, single dose, usually less than 24 hours (or no more than one week).
- **Subacute tests**: repeated doses for up to but no more than one month.
- **Sub-chronic tests**: repeated doses from one to three months, or possibly up to 10% of the life span of the species studied.
- **Chronic tests**: repeated doses for more than three months, or at least 10% (and typically 50% or more) of the life span of the species.

For example, exposure to a single dose of benzene can result in the depression of the CNS, while repeated doses over a longer period can result in leukemia.

After careful analysis of laboratory data, parameters are established by the group or organization conducting the testing or enforcing the standards, i.e., limits of exposure. A summary of the major parameters and terms used in risk assessment is as follows:

- **DOSE:** the amount of chemical toxicant entering the body through any of the three main exposure routes.
- **TLV:** threshold limit value, refers to ambient airborne concentrations of a substance of interest and represents a condition under which it is believed that nearly all workers may be repeatedly exposed without suffering adverse effect. This limit is "advised or recommended" (not enforceable) by the American Conference of Government Industrial Hygienists (ACGIH) and is based on epidemiological studies or testing of laboratory animals. Concentration units may be expressed in one of two ways: mass/mass, i.e., ppm or mg/kg; or mass/volume, i.e., $mg/m^3$.

- **TLV-TWA:** threshold limit value–time-weighted average for an 8-hour workday over a 40-hour work week to which any and all workers may be subjected without suffering adverse effect. TWA is a suffix that can be added to PEL or REL values (as listed below).
- **TLV-STEL:** Similar to a TLV, except as a short-term exposure limit, it is the concentration to which workers can be continuously exposed for short periods of time without suffering from irritation, chronic or irreversible tissue damage, or narcosis to a sufficient degree to increase the likelihood of accidental injury, impair self-rescue, or materially reduce work efficiency. The maximum exposure time is usually taken to be 15 minutes. STEL is a suffix that can be added to PEL or REL values as listed below.
- **TLV-C:** See comment above, except as a ceiling value, never to be exceeded even instantaneously during any part of a work day.
- **IDLH:** Immediately dangerous to life and health, never to be exceeded, somewhat similar to a TLV-C; it is a concentration from which a worker can escape in 15 minutes or less without suffering any escape-impairing symptoms or irreversible health effects. This limit is established by NIOSH.
- **PEL:** permissible exposure limit is an ambient, airborne concentration of a substance of interest that workers can be regularly subjected to without suffering any adverse effects. An 8-hour work day and a 40-hour work week are assumed (and may be confirmed by attaching the "TWA" as a dashed suffix). This concentration has been adopted by OSHA (Occupational Safety and Health Administration, which is a branch of the U.S. Department of Labor) and is based largely on data furnished by the ACGIH. The initial listing of PELs was made in the Z-Tables of Title 29, Code of Federal Regulations, Part 1910.1000, published in the *Federal Register* on January 19, 1989. The suffixes TWA, STEL, or C may be added as necessary with the same meanings and definitions referenced above in TLVs.
- **REL:** relative exposure limit is still another ambient airborne concentration of a substance of interest that has been researched, developed, and recommended by NIOSH (National Institutes of Occupational Health and Safety), based on a 10-hour workday. RELs are often, but not always, adopted by OSHA. The suffixes TWA, STEL, or C may be added as necessary with the same meanings and definitions referenced above in TLVs.
- **NOAEL:** no observable adverse effect level.
- **LOAEL:** lowest observable adverse level.
- **ADI:** acceptable daily intake usually applies to a food additive over a lifetime, but may be used to describe any chemical substance. This is calculated as the NOAEL/safety factor, where the safety factor is usually 1000. The safety factor takes into account extrapolations from animal studies to human subjects and differences in metabolism among individuals in humans.
- **LD$_{50}$ or LC$_{50}$:** also known as the "median lethal dose," this is the lethal dose by oral ingestion or dermal contact for 50% of a population of a given species; the LC$_{50}$ value (a.k.a. "median lethal concentration") is the lethal concentration by inhalation for 50% of a population of a given species. LD$_{50}$ values are usually expressed in terms of mg of toxicant per kg of body

weight. The poison is administered as a one-time dose. This is a measure of *acute* toxicity. The $LC_{50}$ value can refer to a toxicant dispersed in air or dissolved in water and inhaled or ingested, respectively. To calculate the effective lethal dose for a subject in question, the $LD_{50}$ value is by the subject's mass in kg, i.e., lethal dose = $LD_{50}(mg/kg) \times Mass(kg)$.

- **Action Level:** this is an 8-hour TWA concentration for which there is a risk of having more than 5% of the employee workdays involving an exposure at a concentration greater than the relevant TLV-TWA, PEL-TWA, or REL-TWA.
- **Neurotoxin:** a chemical substance toxic to the CNS.
- **Hemotoxin:** a chemical substance toxic to the blood.
- **Hepatotoxin:** a chemical substance toxic to the liver.

As previously mentioned, epidemiological studies of toxins on humans are often not available. When laboratory studies of animals have been conducted, the most difficult aspect is interspecies variation of test results. Which species should be used as a test population? Which species are closest to humans? Many species react quite differently when exposed to the same toxin. It has also been recently established that gender differences in humans determine degrees and limits of toxicity.

For example, the chemical dioxin, an endocrine-disrupting chemical or EDC (this topic is discussed in detail in Section 15.8) is found as an unwanted by-product and contaminant in the herbicide Agent Orange used in Vietnam, as a defoliating agent, and has been extensively studied. Pure dioxin is a single compound whose molecular structure is shown in Figure 15.1. In its pure form, it is correctly identified and named as 1, 4 dioxin, $C_4H_4O_2$ (not to be confused with 1,4 dioxane, $C_4H_8O_2$ discussed in Section 15.6).

However, the term dioxin is also commonly used to describe a family or class of compounds containing the basic dioxin structure. (Recall that dioxins include three subclasses of compounds, polychlorinated dibenzo-p-dioxins, polychlorinated dibenzofurans, and PCBs, discussed in Chapter 13.6.) The individual dioxin compound most commonly encountered and studied is 2,3,7,8-tetrachlorodibenzo-p-dioxin (TCDD) with a molecular formula of $C_{12}H_4Cl_4O_2$, shown in Figure 15.2.

**FIGURE 15.1**   Molecular structure of 1,4 dioxin.

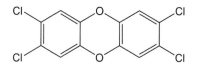

**FIGURE 15.2**   Molecular structure of TCCD.

**TABLE 15.4**
**Parameters to Measure Toxicity**

| Substance | $LC_{50}$ | Exposure Limit | STEL | IDLH |
|---|---|---|---|---|
| Ammonia | 4000 ppm | 50 ppm (OHSA-PEL) | 35 ppm (NIOSH) | 300 ppm |
| | | 25 ppm (NIOSH-REL) | | |
| Carbon monoxide | 3760 ppm | 35 ppm (NIOSH-REL) | | 1200 |
| | | 200 ppm (NIOSH-C) | | |
| | | 50 ppm (OSHA-PEL) | | |
| Chlorine | 293 ppm | 0.5 ppm (NIOSH-REL) | | 10 ppm |
| | | 1.0 ppm (OSHA-PEL) | | |
| Hydrogen cyanide | | 10 ppm (OSHA-Skin) | 4.7 ppm (NIOSH –Skin) | 2 ppm |
| Phosgene | 800 ppm | 0.1 ppm (NIOSH-REL) | 0.03 ppm (NIOSH) | 4 ppm |
| | | 0.1 ppm (OSHA-Skin) | | |
| | | 0.1 ppm (ACGIH-TLV) | | |

Note that the $LD_{50}$ for TCDD in guinea pigs is 0.0016 mg/kg, while the $LD_{50}$ in hamsters is approximately 3 mg/kg from Table 15.1B. The variation is significant. The testing is done to assess the risk in humans. Clearly, humans are neither guinea pigs nor hamsters.

Furthermore, this information is only about lethal doses. Other classification modes of action for toxins—skin irritants, allergens, hemotoxins, neurotoxins, endotoxins, mutagens, teratogens, and carcinogens—have not been scientifically evaluated.

Table 15.4 shows toxicity measurements of selected common gases and vapors. It is interesting to note that a complete set of data for each substance is not always available. It is also interesting to note the differences in exposure limits among various agencies. For a more complete guide to toxic substances, their exposure limits by category, and their health effects including target organs, consult the NIOSH Pocket Guide to Hazardous Substances available for order on DVD (Tel: 1-800-356-4674), hard copy (Tel: 1-800-232-4636), or at the NISOH website: http://www.cdc.gov/niosh), or in direct electronic or DVD format available for download at the NIOSH website: http://www.cdc.gov/niosh/npg/npg.html.

## 15.4   THE DOSE–RESPONSE RELATIONSHIP

A large portion of toxicology deals with the concept of dose of toxin and its concomitant response in the subject, i.e., dose–response relationships or curves. This knowledge will assist regulatory agencies, such as OSHA and EPA, in enforcing toxicity standards and regulations in an effective manner. It might seem intuitive to assume that increasing the dose of a poison should automatically increase the percentage of respondents in a given population and probably in a linear fashion. This is not always the case, as the effects are not always predictable, much less the target organs or body systems affected. Consider the following five "Dose–Response" relationships.

Figure 15.3 is a graph that shows two common S-shaped dose–response profiles or relationships for two different toxicants, A and B. Note that toxicant B has a lower threshold limit at 5 mg/kg than B at 10 mg/kg. The threshold limit marks the point at

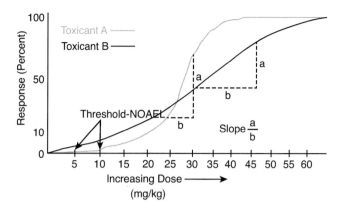

**FIGURE 15.3**  Standard S-shaped dose–response curve for two toxicants, A (in grey) and B (in black).

which the smallest dose of the toxin produces a noticeable effect. Below this threshold limit, NOAEL has been detected in any member of a particular test population. As the dose is increased, however, the percent of members in the test population showing symptoms begins to increase, until a point is reached where all members (i.e., 100%) show symptoms. Also note that until a dose of 25 mg/kg is reached, B is a more toxic substance. At 25 mg/kg, A and B intersect and are equal in toxicity. Thereafter, note that A has a steeper slope value and thus has a greater sensitivity of response, until a dose of about 70 mg/kg is reached, where the two lines converge.

Figure 15.4 shows two toxicants, A and B, in linear, parallel relationships, meaning that doubling the dose doubles the percent response of symptoms in the species. Note that toxin B has no threshold value, while A has a threshold value of about 600 mg/kg. B produces an earlier onset of symptoms.

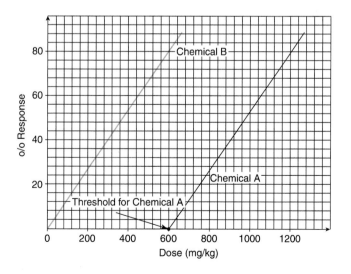

**FIGURE 15.4**  Two toxicants, A and B, different threshold values with linear responses.

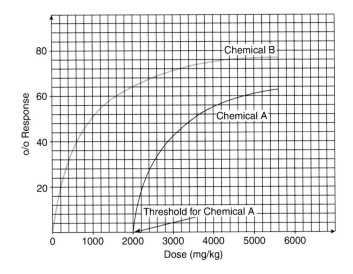

**FIGURE 15.5**  Two toxicants, A and B, different threshold values with nonlinear responses.

Figure 15.5 shows two toxicants again, A and B, but in nonlinear and fairly congruent relationships. Note that again B has no threshold value, but A has a threshold value of about 2000 mg/kg. The maximum percent response as the dose is increased indefinitely is about 76% for B and about 64% for A. This implies that about 24% of the test population are immune to toxicant B, while 36% are immune to toxicant A.

Figure 15.6 shows two toxicants, X and Y, with very different relationships. Toxin Y is more toxic at low doses and in a nonlinear or more rapid rate, while toxin X is

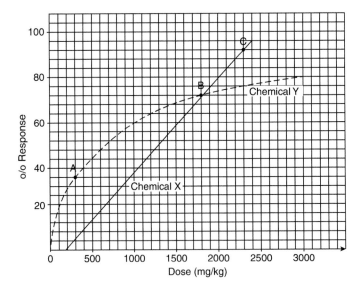

**FIGURE 15.6**  Two toxicants, X (straight line) and Y (dashed Line), with linear and nonlinear intersection.

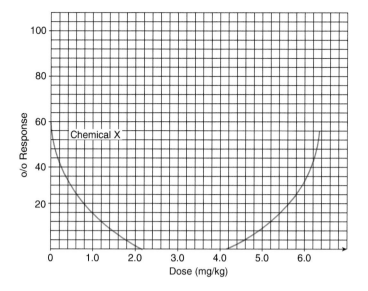

**FIGURE 15.7**   Inverted/U-type counter intuitive response.

more toxic at higher doses and in a linear fashion. The point of intersection, B, marks the dose (i.e., ~1800 mg/kg) at which both toxins produce the same response rate of about 72%. Above this dose, X is more toxic, while Y plateaus and never produces more than an 80% response rate.

Figure 15.7 shows a plot of a dose–response relationship that is counterintuitive with U-shaped response. It shows a chemical toxicant X that shows detectable symptoms in a large number of the population at very low doses, i.e., it has no threshold value, then has a *declining* response rate with increasing dose, in a nonlinear manner until about 2.0 mg/kg. However, at higher doses, the response rate starts to climb again. The shape is a large U. This chemical has no safe level and shows a region where a slightly higher dose, e.g., 4.5 mg/kg, may be safer than a very low dose, e.g., 1.0 mg/kg or less. Endocrine disruptors, discussed in Chapter 7, such as the pesticide atrazine, are a class of chemical substances that exhibit this type of behavior.

## 15.5   MULTIPLE CHEMICAL SENSITIVITY (MCS) SYNDROME

In the late 1980s, a new condition began to emerge known as **Multiple Chemical Sensitivity (MCS)**. MCS is a chronic condition marked by heightened sensitivity to multiple different chemical substances and/or irritants at or below previously tolerated concentrations of exposure. Sensitivity to odors is often accompanied by food and drug intolerances, photosensitivity to sunlight, hypersensitivity to touch, heat, and/or cold, loud noises. MCS may be caused by acute exposure or repeated, chronic, low-concentration exposure to environmental, chemical or nonchemical agents. While some reactions resemble symptoms commonly associated with an allergy, it is important to understand that reactions to chemical agents are more serious and can result in neurotoxic effects, which can impair normal functioning. Except for MCS victims themselves, few people recognize that reactions to environmental triggers may become so debilitating that drastic changes in lifestyle may have to be made just to survive. The picture that emerges that most experts in the field agree on is that environmental pollutants tend to bioaccumulate into a net body burden that triggers an immune response, when the threshold is crossed. That threshold is unique and different for every person.

Reported, documentable reactions have included symptoms such as headache, nausea, dizziness, fatigue, burning of the eyes, nose, and throat, gastrointestinal problems, musculoskeletal pain, sleep disorders, seizures, impaired balance, loss of memory or concentration, and even cognitive dysfunction. Many doctors now believe that MCS is actually a manifestation of an **autoimmune disorder**, e.g., irritable bowel syndrome, lupus, rheumatoid arthritis, or multiple sclerosis. Over 30 other terms have been used to describe this condition in the past, including environmental illness, environmental irritant syndrome, chronic fatigue syndrome, sick building syndrome, cacosmia, and toxic response syndrome. While initially dismissed as an exaggerated form of a temporary allergic reaction, it is now recognized as a legitimate, documentable disease all over the world, including the United States. Triggers are wide ranging and include exposure to gasoline vapors and hydrocarbon fuels, pesticides, household cleaning agents, food additives, perfumes, vehicle exhaust, moth balls/flakes, flame retardants in curtains and sofas, building materials and synthetic carpets or clothing which off-gas formaldehyde or other VOCs, and other "non-specific" air pollutants. The best treatment for MCS is the avoidance of the offending substances, by eliminating one substance at a time. This has proven to be the only permanent solution, with some victims experiencing only partial recovery. MCS is one reason why both indoor and outdoor air quality is so important.

## 15.6  PESTICIDES: TWO EXAMPLES—ATRAZINE, ARSENIC

An abundance of pesticides, i.e., herbicides, insecticides, fungicides, exists in the marketplace today, some with a longer history or more notoriety than others. Chlorpyrifos (insecticide), dicamba (herbicide), glyphosate (herbicide), atrazine (herbicide), and arsenic, to name just a few, are in the marketplace as of this publication. Chlorpyrifos will be prohibited for use at the end of 2020 in California and may be on the way out in other states as well. It is already illegal to either sell or use it in the European Union through its R.E.A.C.H. (Registration, Evaluation, and Authorization, and Restriction of Chemicals) program.

The equivalent of the USEPA in the European Union is ECHA, i.e., The European Chemical Agency. ECHA administers the REACH program, under which all chemicals manufactured, imported, or otherwise used in end products for sale in the European Union must be registered in a master inventory. The EPA has a similar program through TSCA and RCRA (Resource Conservations and Recovery Act), both enacted in 1976. One chief difference between the USEPA and the ECHA is the number of chemicals assessed and the method by which they are evaluated for risk or toxicity and determined to be safe.

EPA has just reauthorized a new 5-year registration of two products containing dicamba as of November 2020 in the United States, despite numerous complaints by farmers about its drifting to unwanted fields.

According to a series of articles published in Chemical and Engineering News during the past 5 years, glyphosate continues to be a highly controversial herbicide, used in Monsanto's Roundup, now produced by Bayer. In 2016, a UN expert panel, the IARC, and the WHO each declared that glyphosate is a probable human carcinogen and teratogen and toxic to many endangered species of fish and wildlife. There is strong evidence that it impacts the guts of honeybees, making them more susceptible to bacterial infections and ultimate death. Despite numerous studies to the contrary and objections by many environmental watchdog groups, the USEPA has recently greenlighted its use and cites it is not a carcinogen, although it is still accepting complaints as of this publication. Meanwhile, as of May 2019, 13,000 lawsuits have been filed against Monsanto and juries in two cases have awarded plaintiffs with cancer $80 million each. The other cases are still pending.

Glyphosate is currently approved for use in the European Union but only until December 15, 2022. Austria has already banned its use as of July 2016.

### 15.6.1 ATRAZINE

In a 2010 U.C. Berkley CA research study conducted by biologist, Tyrone Hayes, 40 African clawed frogs were exposed to 2.5 ppb of atrazine in water solution for 3 years—a level below the 3.0 ppb level allowed by the US EPA for drinking water. (Note: 50 ppb is roughly equivalent to one drop of water in an Olympic swimming pool.)

As a result, 30 of the 40 frogs were chemically castrated; 75% were incapable of reproducing. They were missing testosterone and the ability to make sperm.

Four of the 40 actually turned into females, eventually mating with other males and producing viable eggs, despite being genetically male. All of their offspring were male.

Atrazine is a prime example of an **endocrine-disrupting chemical, an EDC** (EDCs are discussed in the next section). Some 80 million pounds of atrazine are applied annually in the United States on corn and sorghum to control weeds and increase crop yield. More than 33 million Americans estimated to be exposed as of 2009.

Subsequent studies have shown that frogs exposed to atrazine at concentrations of even 0.1 ppb—30 times lower than the EPA limit—show comparable results.

The European Union, through its R.E.A.C.H. program, banned the use of atrazine in October 2003, the same time EPA approved it in the United States. It has been detected in **both** US drinking and bottled water. After a 7-year-long research study, EPA has approved and reauthorized the use of Atrazine once again as of November 2020 but with some restrictions on its use going forth.

For comparison, it is interesting to note that in the area of cosmetic products, the EU's ECHA has banned 1328 chemical substances (e.g., coal tar, parabens, hydroquinone, formaldehyde, and triclosan), while the United States has banned only 30. One has to wonder if this is causally linked to the "FIVE IMMEDIATE CONCLUSIONS" deduced from the 2017 Lancet Commission Report presented at the beginning of this chapter.

## 15.6.2 THE PERSISTENCE OF ARSENIC—A CROSSOVER CHEMICAL—A PESTICIDE AND MORE

Arsenic occurs naturally and ubiquitously in the Earth crust. It contaminates water in regions all over the world, including the United States. The WHO has called the arsenic contamination in Bangladesh "the largest mass poisoning of a population in history." In European soils, the average concentration of arsenic is 7.0 mg/kg (Salminen, et al, 2005), and in US soils, it is 5.6 mg/kg (Gustavsson et al, 2001).

Arsenic is a metalloid and has a complex chemistry. In aquatic systems, it can originate from both natural and anthropogenic sources. Sources include volcanic eruptions, weathering of rock, burning of fossil fuels, application of pesticides, dust particles from mining operations and metal refining processes, wood preservatives (pressure-treated lumber), cement works, electronic industries, ammunition factories, and pyrotechnics. Arsenic is usually found in one of four oxidation states: $As^{5+}$ or As(V) as arsenate; $As^{3+}$ or As(III) as arsenite; $As^o$ as arsenic; and $As^{3-}$ or As(III$^-$) as arsenide. $AsH_3$ is arsine, a highly toxic gas and a hemotoxin. In the crust or lithosphere of the Earth, arsenic is commonly found as an arsenide of copper, iron, or nickel. It is also found as arsenic oxide or arsenic sulfide. Arsenic(III) oxide is generated from copper smelting.

Like mercury, arsenic exists in both organic and inorganic forms. Unlike mercury, the inorganic variety is more toxic than the organic. In the inorganic form, As(III) compounds, i.e., arsenites ($AsO_3^{3-}$), are more toxic than As(V) compounds, i.e., arsenates ($AsO_4^{3-}$). Aerobic conditions favor the formation of As(V) compounds. Thus, water contaminated with arsenic aerated from a sprinkler has a higher concentration of As(V) than water flowing naturally. In research conducted in Sardinia, Italy, it was no surprise to learn that rice grown in continuously flooded waters was found to have much higher As(III) levels ranging from 95 to 235 µg per kg, than rice grown in fields with sprinkled water, which ranged from 1.3 to 5.1 µg per kg (Spanu et al. 2012).

Drinking water tends to be the single, largest source of arsenic. Most of this is from natural sources, but some may be due to anthropogenic sources, such as pressure-treated lumber and pesticides discussed later in this section. Concentrations of arsenic often exceed US EPA standards for drinking water as well as in common foods such as rice. Rice has been described as a natural "sponge" and can accumulate toxins, particularly heavy metals and metalloids. In November 2012, Consumer Reports (CR) released an extensive report that showed of 223 rice and rice products tested, most exceeded the EPA limit or standard on inorganic arsenic in drinking water of 10 parts per billion (10 ppb). Some products tested as high as 270 ppb.[1] The following year, US FDA tested 1300 samples of rice and rice products. It released its own report showing similar results: average levels of inorganic arsenic ranged from 0.1 to 7.2 ug per serving of rice or rice products. For an average serving of 45 g rice, this equates to 0.002–0.16 mg of arsenic per serving.

---

[1] In January 2013, CR also released a report on arsenic in fruit juices. It showed that 25% of the 31 samples tested exceeded the 5 ppb concentration established by EPA for bottled water, and 10% exceeded the 10 ppb concentration established for drinking water. The source of the arsenic is unknown and may be natural groundwater, or contamination from the use of pressure-treated lumber or pesticide spraying of trees.

However, neither the United States nor the European Union has established legal limits for arsenic content in food. In 2014, the WHO recommended a maximum arsenic concentration of 0.20 mg per kg rice. Just recently, the European Commission, the executive arm of the EU government, subsequently proposed the following limits: 0.2 mg/kg white rice; 0.25 mg/kg for brown rice; 0.3 mg/kg puffed rice; and 0.1 mg/kg for rice products intended for children. The European Union expects that these limits will be adopted in 2016.

Meanwhile, the US FDA has been encouraging drug makers to remove more than 100 arsenic-based veterinary drugs from the marketplace. Such drugs are given to poultry to prevent disease. FDA announced in the spring of 2015 that one manufacturer, Zoetis, will stop selling nitarsone later in the year. Urvashi Rangan, executive director of Consumer Reports' Food Safety & Sustainability Center, remarked: "Because poultry won't be fed arsenic, a known carcinogen, the spread of arsenic into the food supply will be curtailed. Poultry manure used for fertilizer will contain less arsenic and won't contribute to contamination of crops and the environment."

Regardless of the source, infants and children exposed to inorganic arsenic are put at increased risk of getting type-2 diabetes, heart disease, and certain types of cancers later in life. Interestingly, in contrast to rice or fruit juice, eating a 2-pound lobster, an established bottom feeder, can boost organic arsenic blood concentration to dangerous levels. But because it is in organic form, the human liver and kidneys will metabolize arsenic to safe concentrations within 72 hours and excrete it.

One anthropogenic source of arsenic that deserves special attention is **pesticides**. Pesticides include herbicides, insecticides, fungicides, and rodenticides. The EPA currently lists 14 arsenical pesticide compounds that it monitors, e.g., lead arsenate and copper arsenite, sold under various trade names. While these are useful in controlling disease and protecting food supply, mounting scientific evidence exists that shows arsenic is more harmful to human health, especially children, than originally thought. Pesticides can be inhaled either during or after application, ingested if fruits or vegetables are not properly washed, or can be absorbed through the skin by touching surfaces that contain surface residues from having been treated. Pesticide exposure can cause both acute and chronic health effects. Acute health effects include coughing, shortness of breath, nausea, vomiting, eye irritation, and headaches. Chronic health problems, based on over 40 years of research, include a wide range of conditions associated with type 2 diabetes, cardiovascular disease, blood disorders, and certain cancers (lung, urinary bladder, skin, and kidney). Of course, children are at greater risk because of their smaller weight (mass) as well as the fact that they are still in their developmental stages. The best strategy to treat arsenic poisoning remains chelation therapy or kidney dialysis.

Another anthropogenic source of arsenic that deserves attention, especially in the urban environment, is pressure-treated lumber used in outdoor construction, such as decks and outdoor playground equipment in school yards and public playgrounds. Copper–chromated–arsenic (Cu-Cr-As or CCA) is an excellent fungicide (Cu) and insecticide (As) and is used to impregnate wood as a long-lasting preservative. The purpose of chromium is to "fix" the copper and arsenic to the wood fibers. Many studies exist and have shown that arsenic leaches out of CCA-treated wood into the surrounding soils over time. (Arsenic is detected in soil samples by atomic

- Argentina (<1 to 2550 ug/L)
- Bangladesh (<10 to >2500 ug/L)
- Chile (600-800 ug/L)
- China (<50 to 4400 ug/L)
- India (<10 to >800 ug/L)
- Mexico (5 to 43 ug/L)
- Taiwan (<1 to >3000 ug/L)
- United States (<1 to >3100 ug/L)
- Viet Nam (<0.1 to 810 ug/L)

**FIGURE 15.8**   Countries with high, spot levels of arsenic in their drinking water. (Data from M.E. Naujokas, et al. *Environmental Health Perspectives* 121:295.)

absorption spectrometry.) A 2004 Canadian study found that children who played on CCA-treated wooden playground equipment did indeed have higher amounts of arsenic on their hands than children who played on playground equipment made from other materials. However, the study also found that the overall exposure was lower than typically ingested from food and water sources. A 2010 Canadian study found no significant difference in arsenic concentrations in the urine and saliva of children playing on CCA-treated wooden playgrounds compared with children playing on other playgrounds (Figure 15.8).

## 15.7   ENDOCRINE-DISRUPTING CHEMICALS (EDCS): BPA, DEHP, AND DOP

Throughout the human body are various receptor cells or "locks" waiting to be turned on or off at the right moment by hormones or "keys." Turning these receptor cells on or off helps regulate all biological processes or "systems" in the body. These systems are known collectively as the endocrine system and consist of a complex network of glands. Endocrine disruptors, then, are a class of often synthetic, chemical substances, which interfere with, or in extreme cases, block, the proper, intended functioning of cells. Disruption in the natural timing and sequence of the cell functions in the endocrine system may have negative effects ranging from prostate cancer, breast cancer, and testicular cancer, all three of which are on the rise, to increased rates of obesity, diabetes, and infertility (low sperm counts). In fact, several studies indicate that male sperm counts in the United States have declined 30% over the last 50 years. The interested reader can consult research studies conducted by Frederick vom Saal, a pioneer and leading expert in this field.

The group of molecules identified as endocrine disruptors is quite heterogeneous and includes synthetic chemicals used as industrial solvents, lubricants, plasticizers, and pesticides, and their by-products. According to the National Institute of Environmental Health and Sciences (NIEHS), a partial list or examples of **EDCs** currently include BPA, dioxins, PBDEs, PCBs, phthalates, perchlorate, perfluoroalkyl and polyfluoroalkyl substances (PFAS), and triclosan.

For the reader who may not be familiar with the source or extent of use of some of the EDCs mentioned above, Table 15.5 indicates some more commonly encountered applications.

**TABLE 15.5**

**Endocrine Disruptor Chemicals and Where They Are Found**

| EDC | Applications |
|---|---|
| BPA | A monomer in the production of polycarbonate plastic used in dental sealants, beverage bottles (recently phased out), and transfer agent in cash register receipts |
| Dioxins | Unwanted by-products/contaminants in the production of Agent Orange |
| PBDEs | Flame retardants in curtains, mattresses, quilts; electrical insulation |
| PCBs | Insulating fluids used in electrical transformers |
| Phthalates | Plasticizers used to soften PVC plastic; vapor pressure suppressant in colognes and perfumes |
| Perchlorate | Used in the production of ammunition and weapons as explosives |
| PFAS | Used in the manufacturing of Teflon; also as hydrophobic and lipophobic coatings in take-out food containers |
| Triclosan | An antibacterial and antifungal agent present in soaps, toothpaste, detergents, leading to antibiotic resistance. |

Three examples of endocrine disruptors discussed in this section are BPA used, until recently, to manufacture polycarbonate plastic for beverage containers, and two phthalate esters, di (2-ethylhexyl) phthalate (DEHP) and di-n-octyl phthalate (DOP) used as plasticizers. (PFAS and 1,4 Dioxane (not classified as an EDC) will be discussed in Section 15.8 under Emerging Contaminants.) Their molecular structures are shown below in Figures 15.9 and 15.10. BPA is drawn in Figure 15.11, and again redrawn in Figure 15.12, along with another example of an endocrine disruptor, DES (diethylstilbestrol), and 17 beta estradiol, the most common human estrogen, to facilitate easier comparison of molecular structures.

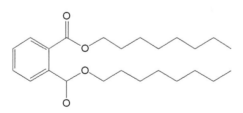

**FIGURE 15.9**   Molecular structure of di-n-octyl phthalate or DOP.

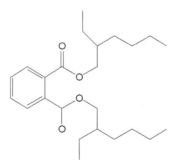

**FIGURE 15.10**   Molecular structure of di (2-ethylhexyl) phthalate or DEHP.

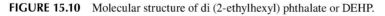

**FIGURE 15.11**   Molecular structure of bisphenol A or BPA.

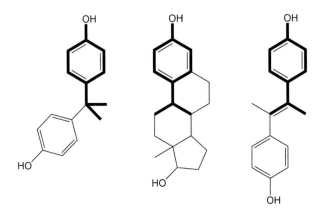

**FIGURE 15.12**   A comparison of three molecular structures.

From left to right, a comparison is made of the similarities in molecular structures of BPA (far left), 17 beta estradiol, the most common human estrogen (middle), and DES, diethylstilbestrol (far right). DES is an artificial hormone synthesized in 1938 and was given to women from 1940 to 1971 to reduce pregnancy complications or miscarriages, later withdrawn when it was shown to be carcinogenic.

These molecules are also known as estrogen mimics, since they imitate the chemical properties and reactions of natural estrogens within the human body. Note the particular similarity in molecular structure between BPA, as well as DES, and 17 Beta Estradiol, the most common human estrogen. They can interfere with the activity of natural hormones, like estrogens. Hormones act as chemical messengers traveling throughout the network of glands in the body. They control metabolism, reproduction, growth, and other functions. BPA has been found to induce epigenetic changes. Other chemicals in the class of endocrine disruptors include dioxins and its three subclasses (discussed in Section 13.7.3 and whose molecular structures are provided in Figure 13.6), polybrominated diphenyl ethers (or PBDEs, used as flame retardants), atrazine and pentachlorophenol (organochlorines, used as pesticides), and perfluoro-octanoic acid (or PFOAs, used in the manufacturing of Teflon-coated products). Use of DDT and PCBs has been banned in the United States since 1978, and PFOA is being phased out.

Here are some summary facts for each.

BPA:

- Originally synthesized by a Russian chemist in 1888, by reacting acetone with phenol.

- BPA is subsequently reacted with phosgene or carbonyl chloride ($COCl_2$) to produce polycarbonate, a hard durable plastic, used to make DVDs, cell phones, computer casings, safety glasses and lenses, sports equipment, automobile parts, and water bottles.
- BPA can also react with epichlorohydrin ($C_3H_5OCl$) to form an epoxy resin, used for food and drink can liners, as well as paints, coatings, and adhesives.
- Global demand and production are over 12 billion pounds annually—74% to make polycarbonate and 20% to make epoxy liners for food cans.
- Multiple studies indicate that BPA has been linked to obesity, diabetes, fertility disorders in both men and women, and cognitive and behavioral disorders like ADHD.
- According to Johns Hopkins Bloomberg School of Public Health in 2013, detectable levels of BPA have been found in 95% of the adult population in the United States.
- The good news to report is that blood concentration of BPA seems to be dropping in the US population since 2000, based on CDC's National Health & Nutrition Examination Surveys conducted in 2005–2006, 2007–2008, and 2009–2010.
- Banned for use in baby bottles in China effective 2011. See below for actions taken in the European Union.
- Legislation passed between 2009 and 2011 banning BPA use in *certain* products for *children* in CT, MD, MN, NY, VT, WA, and WI in the United States.
- Specialty papermaker, Appvion, has recently designed a process that uses vitamin C (ascorbic acid) instead of BPA as the transfer agent to enable thermal printing of heated cash register receipts.

DEHP and DOP:

- DEHP is one of the most abundantly manufactured and used plasticizers, originally intended as a softening agent in PVC.
- DEHP leaching qualities first brought to recent public attention when PVC intravenous bags used in hospital blood transfusions were found to contain 800 times the safe concentration of DEHP.
- The sale of all phthalates in toys for children was banned by the CPSC as of February 10, 2009 in the United States. The city of San Francisco imposed an earlier ban on the manufacturing, sale, and distribution of all child care articles and toys containing most phthalates (and BPA) in June of 2006.
- DEHP is known to cause liver cancer and impair sperm counts.
- On July 5, 2015, the European Union through its R.E.A.C.H. program, banned the use of DEHP and other phthalates such as, DBP (di butyl phthalate) and BBP (benzyl butyl phthalate] in *all* child care products and toys and restricted the use of three other phthalates (DIDP, DNOP, and DINP) in products that children can put in their mouths.
- However, most other countries do not prohibit or restrict the use of phthalates as plasticizers.

The problem with both of these EDCs is that they have a strong potential for leaching out of their solid plastic matrix into the food or beverage they contain and thereby getting ingested. Since they are fat-soluble, they tend to leach out more into foods that contain solubilized fats and oils, and they bioaccumulate into lipids and adipose tissue of animals.

The reason behind this is as follows. In the case of BPA, not every single molecule reacts with phosgene or epichlorohydrin to form polycarbonate plastic or resin. Unreacted molecules trapped in the plastic matrix gradually migrate and leach out over time under ambient conditions. Heat and sunlight accelerate this migration. The consumer of ambient air is the ultimate sink.

DEHP, DOP, and other phthalate esters are added to the PVC matrix to help soften the plastic and make it more pliable and are called **plasticizers**. They act as a spacer molecule or internal lubricant to help the long-chained PVC polymers, often millions of segmer units long, to move, slip, and slide past one another. This is what gives PVC its supple and flexible quality. However, the plasticizer molecules are held in place by the weakest of intermolecular forces, and over time, become fugitives, and migrate out of the plastic matrix, leaving a more brittle matrix behind. Heat and sunlight accelerate this process.

Besides DEHP, there are many other members of the phthalate ester family including di-isononly phthalate (used in garden hoses, shoes, toys); di-n-butyl phthalate (used in cellulose plastics, food wrap, perfumes and colognes, hair spray); di-n-octyl phthalate (used in flooring materials, notebook covers, canvas tarps); and di-n-hexyl phthalate (used in dishwasher baskets, tool handles, flea collars).

Interestingly, most of these EDCs, including BPA, are *more toxic at low doses than at high*. The dose–response curve is counterintuitive. Consider the case of BPA by examining Figure 15.13, showing BPA's toxicity profile and the various doses with their effects, starting with the $LD_{50}$ value at the top. Conclusion: *There really is no safe concentration of exposure.*

## 15.8  EMERGING CONTAMINANTS: TWO EXAMPLES—1,4 DIOXANE, AND PER AND POLY FLUORINATED ALKYL SUBSTANCES (PFAS)

The *Emerging Contaminants Handbook* (Caitlin H. Bell, et al, Editors; CRC Press; Boca Raton, FL; 2019) identifies hundreds of emerging contaminants of immediate concern organized and tabulated by geographic location, date of inclusion, chemical properties, and CASRN, and discusses several of them in some detail. Two of those discussed are presented below. Their mention as emerging contaminants is well deserved because they are of widespread use and prevalence in consumer products and, hence, the environment as well as their remarkably insidious human effects.

### 15.8.1  1,4 Dioxane

1,4 Dioxane is a cyclic ether first synthesized in 1893 and is a multipurpose industrial solvent. See Figure 15.14 for its molecular structure. It becomes unstable at elevated

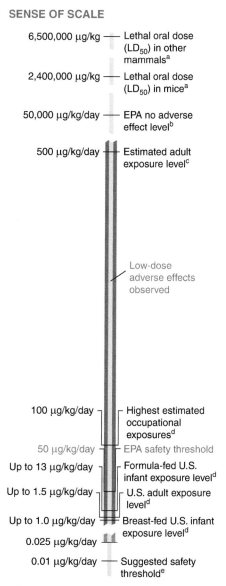

SENSE OF SCALE

| | |
|---|---|
| 6,500,000 μg/kg — | Lethal oral dose (LD$_{50}$) in other mammals[a] |
| 2,400,000 μg/kg — | Lethal oral dose (LD$_{50}$) in mice[a] |
| 50,000 μg/kg/day — | EPA no adverse effect level[b] |
| 500 μg/kg/day — | Estimated adult exposure level[c] |
| | Low-dose adverse effects observed |
| 100 μg/kg/day — | Highest estimated occupational exposures[d] |
| 50 μg/kg/day — | EPA safety threshold |
| Up to 13 μg/kg/day — | Formula-fed U.S. infant exposure level[d] |
| Up to 1.5 μg/kg/day — | U.S. adult exposure level[d] |
| Up to 1.0 μg/kg/day — | Breast-fed U.S. infant exposure level[d] |
| 0.025 μg/kg/day — | |
| 0.01 μg/kg/day — | Suggested safety threshold[e] |

**a** Acute toxicity, where a single dose is lethal to half of the animals treated. **b** EPA estimate of lowest toxicity; the safety threshold is set 1,000 times lower than this value. **c** Daily intake determined by some endocrinology researchers. **d** National Toxicology Program estimated daily intake; formula-fed infant value assumes using polycarbonate baby bottles and formula packaging containing BPA, which have now been phased out. **e** Tolerable daily intake suggested by a group of endocrinologists.

**FIGURE 15.13**    Toxicity Profile of BPA. (Reprinted with the permission from C&E News, 'Sense of Scale,' 2011, pp.14–19, 89 (23). Copyright February 23, 2021, American Chemical Society.)

**FIGURE 15.14**   Molecular structure of 1,4 dioxane.

temperatures and pressures and may form explosive mixtures with prolonged exposure to air or light. Some of its principal uses are a stabilizing agent in chlorinated solvents like 1,1,1 trichloroethane (phased out under the 1995 Montreal Protocol); a primary solvent used in the synthesis of PBDEs (flame retardants) and in degreasing applications; an unwanted by-product in the manufacturing of surfactants used in personal care products (liquid soaps, toothpaste) and in cosmetics; and an unwanted by-product in the synthesis of PET plastic and other polyester resins. It is often found in conjunction with the presence of the surfactant sodium laureth sulfate in liquid soaps. It has been found as an environmental contaminant in soil and drinking water and has been the focus of considerable regulation and treatment since the early 2000s.

In 2013, the USEPA determined that 1,4 dioxane is "likely to be carcinogenic to humans." The ATSDR made this assessment in 2012. The primary routes of exposure are ingestion, inhalation, and dermal contact. It is classified and regulated as a one of several hazardous air pollutants by the EPA (U.S. Clean Air Act of 1999). It is unfortunately completely miscible in water, which makes it difficult to remove. According to a November 9, 2020 article in C & E News, which quoted EWG (Environmental Working Group) as its data source, water monitoring data collected between 2010 and 2015 show more than 7 million people across 27 states, which had utility supplied tap water, had detectable levels of 1,4 dioxane. EPA's recommended or voluntary guidelines, though NOT a primary, enforceable federal standard in drinking water set in 2013, is 0.35 µg/L (or 0.35 ppb). Additionally, a new law (S4389B) calls for a limit of no more than 10 ppm of 1,4 dioxane in cosmetics and no more than 2 ppm in personal care and cleaning products starting December 31, 2022, dropping to 1 ppm after 2023. The toxicity profile is outlined in the above referenced handbook for the interested reader.

The good news is that production of this compound in the United States by BASF stopped as of 2018 when BASF closed its facility in Louisiana. However, BASF continues to import this compound from Germany to supply its US customers.

## 15.8.2  Perfluoroalkyl and Polyfluoroalkyl Substances—PFAS

PFAS are classified as EDCs and have been found in sediment, wildlife, surface waters, groundwater, and ultimately drinking water. They have now been linked with a variety of cancers and reproductive disorders in humans. Known collectively as PFAS, for perfluoroalkyl and polyfluoroalkyl substances, the two subgroups of chemicals in this class comprise roughly 5000 different substances, many of which are isomers and congeners, and all of which are environmentally persistent and have been around since the 1940s. They are often referred to as "forever chemicals." PFAS

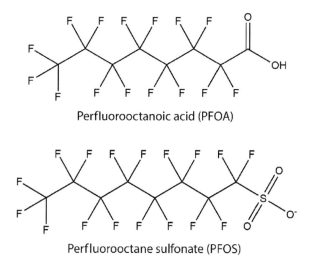

Perfluorooctanoic acid (PFOA)

Perfluorooctane sulfonate (PFOS)

**FIGURE 15.15**    Molecular structures of PFOA (top) and PFOS (bottom).

compounds are ubiquitous and are found in a variety of products from take-out or food delivery boxes to nonstick cookware to stain-resistant clothing. They have both lipophobic and hydrophobic properties. PFAS have been detected in more than 1400 communities across 49 states, according to a recent article in Consumer Reports (CR. org; November 2020). One of their original family members was perfluoro-octanoic acid (PFOA) used to bond Teflon (poly tetrafluoro ethylene) to metallic cookware to make it nonstick. Another early member was PFOS or perfluoro-octane sulfonic acid, both shown below in Figure 15.15, A and B, respectively. These two combined substances are the most studied since the 1990s and believed to be the most dangerous of the group. The EPA set voluntary or recommended guidelines of 70 ppt of these two combined in drinking (tap) water. Note that 1 ppt (part per trillion) is roughly equivalent to one grain of sand in an Olympic size swimming pool. The EWG and other environmental groups believe the limit should be set at 1 ppt. The lack of a primary national standard has implications for not just tap water but also for bottled water, which is regulated by the FDA. Other important and relevant facts related to PFAS are as follows:

- Bowing to pressure from watchdog groups, as of January 16, 2020, the EPA added 160 members of the PFAS group to its Toxic Release Inventory.
- As of 2017, Home Depot has banned the sale of its soil and stain-resistant carpets and rugs treated with two specific PFAS—PFOA and PFOS. Also included are those carpets containing triclosan and vinyl chloride.
- Other retainers are following suit in reformulating their chemical policies: Amazon, Walmart, CVS, and Costco.
- California is halting the sale, manufacture, and use of fire-fighting foam that contains PFAS as of 2022 with limited exceptions. Colorado, New Hampshire, New York, and Washington have enacted similar bans.

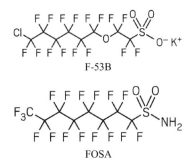

**FIGURE 15.16**    Two random perfluoroalkyl substances—F-53 B and FOSA.

- The US FDA has discovered PFPeA (perfluoropentanoic acid) in chocolate cake and chocolate milk according to unpublished results in a scientific conference in May of 2019. PFPeA is one of many per- and polyfluoroalkyl substances used in nonstick food wrappers and other consumer products for its heat-, stain-, water-resistant properties.
- According to the Pennsylvania Dept. of Health, PFAS have been detected in 95–100% of samples of people's blood in the years 1999–2000 and 2003–2004, although recent monitoring data show the levels of these chemicals in people's blood appear to be declining.
- States such as West Virginia and Ohio have sued DuPont and its 2015 spin-off Chemours for the release of PFOA and PFOS into their states' rivers.
- DuPont has pledged to modify its production of PAFS and end the use of long-chain per- and poly fluoroalkyl substances by the end of 2019 and plans to end the use of short-chain ones as soon as possible thereafter.

Two other lesser known, random, but prevalent PFAS compounds are shown below for comparison in Figure 15.16.

## 15.9 THREE EXAMPLES OF CALCULATIONS OR CONVERSIONS

**Examples 15.1:** Calculations & Evaluations of Air Quality

A solution contains:

- 50 mL benzene ($C_6H_6$) where molar mass = 78 g/mol; vapor pressure = 75 mm Hg; specific gravity = 0.879.
- 25 mL carbon tetrachloride ($CCl_4$) where molar mass =154 g/mol; vapor pressure = 91 mm Hg; specific gravity = 1.595.
- 25 mL trichloroethylene ($C_2HCl_3$) where molar mass = 131.5 g/mol; vapor pressure = 58 mm Hg; specific gravity = 1.455

Assuming an ideal solution and that Raoult's Law is obeyed, what will be the percent concentration of benzene in air at 760 mm Hg pressure at 25°C, assuming the air is saturated with the vapor?

## Solution

Start with Raoult's Law for vapor pressure lowering (see Chapter 3.5, Equation 3.8.
First, recognize that benzene is the solvent since it is present in the greatest proportion.

$$P_{solvent} = X_{solvent}P^\circ{}_{solvent} = X_{benzene}P^\circ{}_{benzene}$$

where $P_{solvent} = P_{benzene}$, i.e., the partial pressure of benzene above the solution.
Next, calculate the number of moles of each component, starting with benzene.

$$\text{Moles of } C_6H_6 = \frac{\text{mass of benzene present}}{\text{molar mass of benzene}} = \frac{\text{volume} \times \text{density}}{\text{molar mass of benzene}}$$

$$= \frac{(50\,\text{mL})(0.879\,\text{g/mL})}{78\,\text{g/mol}} = 0.5635\,\text{moles } C_6H_6$$

$$\text{Moles of } CCl_4 = \frac{\text{mass of carbon tetrachloride present}}{\text{molar mass of carbon tetrachloride}} = \frac{\text{volume} \times \text{density}}{\text{molar mass of carbon tetrachloride}}$$

$$= \frac{(25\,\text{mL})(1.595\,\text{g/mL})}{154\,\text{g/mol}} = 0.2589\,\text{moles } CCl_4$$

$$\text{Moles of } C_2HCl_3 = \frac{\text{mass of trichloroethylene present}}{\text{molar mass of trichloroethylene}} = \frac{\text{volume} \times \text{density}}{\text{molar mass of trichloroethylene}}$$

$$= \frac{(25\,\text{mL})(1.455\,\text{g/mL})}{131.5\,\text{g/mol}} = 0.2766\,\text{moles } C_2HCl_3$$

Now, find the total number of moles present of all components present in solution:
Total Moles = 0.5635 moles $C_6H_6$ + 0.2589 moles $CCl_4$ + 0.2766 moles $C_2HCl_3$ = 1.099 moles
Now, find the mole fraction of benzene in solution, $X_{benzene}$:

$$X_{benzene} = \frac{\text{moles of benzene}}{\text{total moles in solution}} = \frac{0.5635\,\text{moles}}{1.099\,\text{moles}} = 0.5127\,D$$

Next, substitute this value, along with the value for the vapor pressure of pure benzene given as 75 mm Hg, back into Raoult's Law, to find the partial pressure of benzene, $P_{benzene}$, above the solution:

$$P_{benzene} = X_{benzene}P^\circ{}_{benzene} = (0.5127)(75\,\text{mm Hg}) = 38.46\,\text{mm Hg}$$

Finally, to find the percent concentration of benzene vapor in air, determine the ratio of $P_{benzene}$ to the total air pressure given as 760 mm Hg, a consequence of Dalton's Law of Partial Pressures:

$$\%\text{Concentration} = \frac{P_{\text{benzene}}}{\text{Total Pressure}} \times 100 = \frac{38.46 \, \text{mm Hg}}{760 \, \text{mm Hg}} \times 100 = 5.1\%$$

It is interesting to note how large a concentration of benzene in air this answer represents from a health and safety point of view. The OSHA PEL-STEL for benzene is 5.0 ppm. A value of 5.1% is more than 10,000 times as large as the limit of 5.0 ppm.

## Examples 15.2: Calculations & Evaluations of Air Quality

The OSHA PEL-TWA for tetrahydrofuran is 590 mg/m³. Express this concentration in volume units, i.e., in ppm, assuming STP conditions. The molar mass of tetrahydrofuran is 72.1 g/mol.

### Solution

The fundamental principle useful here is the ideal gas law adapted to the information usually provided or immediately available in these situations. Recall that the ideal gas law is

$$PV = nRT = \left(\frac{\text{mass}}{\text{Molar Mass}}\right)RT = \left(\frac{m}{MM}\right)RT$$

This can be modified further to a more useful form as

$$C_{\text{vol}} = \frac{RT}{P(MM)} \times C_{\text{mass}}$$

where $C_{\text{vol}}$ is the concentration in ppm and $C_{\text{mass}}$ is the concentration in mg/m³ of the substance is question, here the tetrahydrofuran. The other quantities are the familiar ones:

$T$ = absolute temperature in kelvin (K) =273K
$R$ = the universal gas constant = 0.0821 L-atm/mol-K
$P$ = pressure in atmospheres = 1.0 atm

Now substitute these values into the above equation and compute the final answer:

$$C_{\text{vol}} = \frac{RT}{P(MM)} \times C_{\text{mass}} = \frac{\left(0.0821\,\text{L atm/K mol}\right)\left(273\,\text{K}\right)}{\left(1.00\,\text{atm}\right)\left(72.1\,\text{g/mol}\right)} \times 590\,\text{mg/m}^3 = 183\,\text{ppm}$$

## Examples 15.3: Calculations & Evaluations of Air Quality

The 8-hour OSHA PEL-TWA and its action level for ethylene oxide are 1.0 and 0.5 ppm, respectively. Compute the action level in mg/m³ at NTP (normal pressure and temperature). The molar mass of ethylene oxide is 44.05 g/mol.

## Solution

Using the same logic in the previous problem, we can adapt and modify the ideal gas law to the following equation:

$$C_{mass} = \frac{P(MM)}{RT} \times C_{vol}$$

In this case, $P = 1.00$ atm, but $T = 298$ K, since NTP indicates normal temperature, which is 25°C or 298 K.

Now substitute the values given into the equation above and compute the final answer:

$$C_{mass} = \frac{P(MM)}{RT} \times C_{vol} = \frac{(1.00\,\text{atm})(44.05\,\text{g/mol})}{(0.0821\,\text{L atm/K mol})(298\,\text{K})} \times (0.5\,\text{ppm})$$

$$C_{mass} = 0.90\,\text{mg/m}^3$$

# Appendix A

Tables A.1–A.5 Distribution of Important Elements in the Four Spheres of the Earth's Environment

---

## TABLE A.1
### Relative Atomic Abundance of Elements in the Atmosphere

| Element | Percent By Number (%) |
|---|---|
| Nitrogen | 78 |
| Oxygen | 21 |
| Argon | 0.93 |
| Carbon Dioxide | 0.039 |

---

## TABLE A.2
### Relative Atomic Abundance of Elements in the Hydrosphere

| Element | Percent by Number (%) |
|---|---|
| Hydrogen | 66.2 |
| Oxygen | 33.2 |
| Chlorine | 0.3 |
| Sodium | 0.3 |

---

## TABLE A.3
### Relative Atomic Abundance of Elements in the Biosphere

| Element | Percent By Number (%) |
|---|---|
| Hydrogen | 49.7 |
| Oxygen | 24.9 |
| Carbon | 24.9 |
| Nitrogen | 0.3 |

## TABLE A.4
## Relative Atomic Abundance of Elements in the Lithosphere

| Element | Percent By Number (%) |
|---|---|
| Oxygen | 61.1 |
| Silicon | 20.4 |
| Aluminum | 6.3 |
| Hydrogen | 2.9 |
| Calcium | 2.1 |
| Sodium | 2.1 |
| Magnesium | 2.0 |
| Iron | 1.5 |
| Potassium | 1.1 |
| Titanium | 0.2 |

## TABLE A.5
## Relative Atomic Abundance of Metals in the Lithosphere

| Metal | Per Cent By Number (%) |
|---|---|
| Aluminum | 8.26 |
| Iron | 5.59 |
| Calcium | 4.12 |
| Sodium | 2.34 |
| Magnesium | 2.31 |
| Potassium | 2.07 |
| Titanium | 0.57 |
| Other | Less than 0.50 |

# Appendix B
## *SDS of Acetone [In Compliance With 1983 OSHA Right-To-Know Law: 29 CFR 1910.1200]*

Safety Data Sheet (SDS)
SDS #: 7
Revision Date: August 31, 2016
Save SDS to Your Library

## B.1  CHEMICAL PRODUCT AND COMPANY IDENTIFICATION

Acetone
Flinn Scientific, Inc. P.O. Box 219, Batavia, IL 60510 (800) 452-1261 Chemtrec
Emergency Phone Number: (800) 424-9344
Signal Word DANGER

## B.2  HAZARDS IDENTIFICATION

Hazard class: Flammable liquids (Category 2). Highly flammable liquid and vapor (H225). Keep away from heat, sparks, open flames, and hot surfaces. No smoking (P210). Hazard class: Eye irritation (Category 2A). Causes serious eye irritation (H319). Hazard class: Specific target organ toxicity, single exposure; Narcotic effects (Category 3). May cause drowsiness or dizziness (H336). Avoid breathing mist, vapors, or spray (P261).

Pictograms

## B.3   COMPOSITION, INFORMATION ON INGREDIENTS

| Component Name | CAS Number | Formula | Formula Weight | Concentration |
|---|---|---|---|---|
| Acetone | 67-64-1 | $CH_3COCH_3$ | 58.08 | |
| Synonyms: Dimethyl ketone, 2-Propanone | | | | |

## B.4   FIRST AID MEASURES

Call a poison center or physician if you feel unwell (P312). If inhaled: Remove victim to fresh air and keep at rest in a position comfortable for breathing (P304+P340). If in eyes: Rinse cautiously with water for several minutes. Remove contact lenses if present and easy to do so. Continue rinsing (P305+P351+P338). If eye irritation persists eyes: Get medical advice or attention (P337+P313). If on the skin: Wash with plenty of water. If swallowed: Rinse mouth. Call a poison center or physician if you feel unwell.

## B.5   FIRE FIGHTING MEASURES

Class IB flammable liquid. A dangerous fire hazard from heat, flame, or strong oxidizers. Flash point: −17°C (CC) Flammable limits: Upper 12.8%; lower 2.6%.

Autoignition temperature: 465°C. When heated to decomposition, may emit toxic fumes. In case of fire: Use a tri-class dry chemical fire extinguisher (P370+P378).

**NFPA Code**
H-1
F-3
R-0

## B.6   ACCIDENTAL RELEASE MEASURES

Remove all ignition sources and ventilate area. Contain the spill with sand or other inert absorbent material and deposit in a sealed bag or container. See Sections B.8 and B.13 for further information.

## B.7   HANDLING AND STORAGE

Flinn Suggested Chemical Storage Pattern: Organic #4. Store with ethers, ketones, and halogenated hydrocarbons. Store in a dedicated flammables cabinet. If a flammables cabinet is not available, store in Flinn Saf-Stor™ can. Keep container tightly closed (P233). Keep cool (P235). Use only in a hood or well-ventilated area (P271). Take precautionary measures against static discharge (P243).

## B.8   EXPOSURE CONTROLS, PERSONAL PROTECTION

Wear protective gloves, protective clothing, and eye protection (P280). Use latex, not nitrile gloves. Wash hands thoroughly after handling (P264). Use only in a hood or well-ventilated area (P271). Exposure guidelines: PEL 1000 ppm (OSHA); TLV 500 ppm, STEL 750 ppm (ACGIH); IDLH 2500 ppm

## B.9   PHYSICAL AND CHEMICAL PROPERTIES

Colorless liquid. Sweet pungent odor like nail polish remover. Soluble: Miscible with water, alcohol, and many other organic solvents.
Boiling point: 56.5°C Density: 0.79 Melting point: −94.6°C Vapor density: 2.00

## B.10   STABILITY AND REACTIVITY

Stable. Potentially explosive reaction with strong oxidizing agents and halogenated compounds. Shelf life: Good, if stored properly.

## B.11   TOXICOLOGICAL INFORMATION

Acute effects: Eye and respiratory tract irritant, dizziness, CNS depression. Chronic effects: Dermatitis. Target organs: Liver, kidneys, CNS, respiratory system.
ORL-RAT $LD_{50}$: 5800 mg/kg IHL-RAT $LC_{50}$: 50,100 mg/m$^3$/8H SKN-RBT $LDL_0$: 20 mL/kg

## B.12   ECOLOGICAL INFORMATION

Data not yet available.

## B.13   DISPOSAL CONSIDERATIONS

Please review all federal, state, and local regulations that may apply before proceeding. Flinn Suggested Disposal Method #18a is one option.

## B.14   TRANSPORT INFORMATION

Shipping name: Acetone. Hazard class: 3, Flammable Liquid. UN number: UN1090.

## B.15   REGULATORY INFORMATION

TSCA-listed, EINECS-listed (200–662-2), RCRA code U002

## B.16   OTHER INFORMATION

This Safety Data Sheet (SDS) is for guidance and is based upon information and tests believed to be reliable. Flinn Scientific, Inc. makes no guarantee of the accuracy or completeness of the data and shall not be liable for any damages relating thereto. The data is offered solely for your consideration, investigation, and verification. The data should not be confused with local, state, federal, or insurance mandates, regulations, or requirements and CONSTITUTE NO WARRANTY. Any use of this data and information must be determined by the science instructor to be in accordance with applicable local, state, or federal laws and regulations. The conditions or methods of handling, storage, use, and disposal of the product(s) described are beyond the control of Flinn Scientific, Inc. and may be beyond our knowledge. FOR THIS AND OTHER REASONS, WE DO NOT ASSUME RESPONSIBILITY AND EXPRESSLY DISCLAIM LIABILITY FOR LOSS, DAMAGE OR EXPENSE ARISING OUT OF OR IN ANY WAY CONNECTED WITH THE HANDLING, STORAGE, USE OR DISPOSAL OF THIS PRODUCT(S).

N.A. = Not available, not all health aspects of this substance have been fully investigated. N/A = Not applicable

Consult your copy of the Flinn Science Catalog/Reference Manual for additional information about laboratory chemicals.

Revision Date: August 31, 2016.

# Appendix C

*Properly Formatted NFPA (704) Diamond Warning Label [in Compliance With 1983 OSHA Right-To-Know Law 29 CFR 1910.1200]*

## Sample Warning Label

**Health**

**4** Extreme hazard—avoid contact or breathing vapor

**3** Severe hazard—use special clothing and masks

**2** Hazardous—use masks or special ventilation

**1** Lightly hazardous—irritating

**0** Normal material

**Flammability**

**4** Extremely dangerous fire and explosion hazard—below 73° F

**3** Fire and explosion hazard at normal temps—below 100° F

**2** Will burn at temps above 100° F

**1** Will burn at temps above 200° F

**0** Will not burn

**Special Notice**

**OXY**—Oxidizing agent

**ACID**—Reacts violently with alkalis

**ALK**—Alkali—reacts violently with acids

**COR**—Corrosive

**W**—Use no water

**P**—Polymerizes

☢—Radioactive

**Reactivity**

**4** Extreme hazard—vacate area in case of fire

**3** Severe explosion hazard

**2** Violent chemical change possible

**1** Unstable if heated

**0** Normally stable

# Appendix D
## Globally Harmonized System (GHS) Symbols

| | | |
|---|---|---|
| • Oxidizers | • Flammables<br>• Self Reactives<br>• Pyrophorics<br>• Self-Heating<br>• Emits Flammable Gas<br>• Organic Peroxides | • Explosives<br>• Self Reactives<br>• Organic Peroxides |
| • Acute Toxicity | • Corrosives | • Gases Under Pressure |
| • Carcinogen | • Irritant | • Environmental Toxicity |

# Appendix E
## *Target of Human Organs by Select Toxicants*

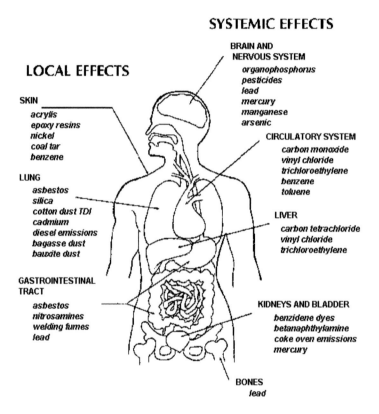

**SYSTEMIC EFFECTS**

**LOCAL EFFECTS**

**SKIN**
acrylis
epoxy resins
nickel
coal tar
benzene

**LUNG**
asbestos
silica
cotton dust TDI
cadmium
diesel emissions
bagasse dust
bauxite dust

**GASTROINTESTINAL TRACT**
asbestos
nitrosamines
welding fumes
lead

**BRAIN AND NERVOUS SYSTEM**
organophosphorus
pesticides
lead
mercury
manganese
arsenic

**CIRCULATORY SYSTEM**
carbon monoxide
vinyl chloride
trichloroethylene
benzene
toluene

**LIVER**
carbon tetrachloride
vinyl chloride
trichloroethylene

**KIDNEYS AND BLADDER**
benzidene dyes
betanaphthylamine
coke oven emissions
mercury

**BONES**
lead

# Bibliography

Acter, T., et al., Evolution of severe acute respiratory syndrome coronavirus 2 (SARS-CoV-2) as coronavirus disease 2019 (COVID-19) pandemic: A global health emergency. (2020). *Sci. Total Environ.* doi:10.1016/j.scitotenv.2020.13924.

Aerosols – Dusts, Fumes and Mists. *The MAK Collection for Occupational health and Safety* 272–292. (2012). Retrieved August 7, 2015 from: http://onlinelibrary.wiley.com/doi/10.1002/3527600418.mb0aeroaere0012/pdf.

Allen, D. and Shonnard, D. *Green Engineering.* Hoboken, NJ: Prentice-Hall. (2002).

American Chemistry Council. *Plastick Packaging Resins.* Retrieved August 7, 2015 from: http://plastics.americanchemistry.com/Plastic-Resin-Codes-PDF.

Anastas, P. and Williamson, T., Editors; *Green Chemistry*; ACS Symposium Series 626. Washington, DC: American Chemical Society. (1996).

Anastas, P.T. and Warner, J.C. *Green Chemistry: Theory and Practice.* Oxford, UK: Oxford University Press. (1998).

Andrews, F. *Thermodynamics.* Hoboken, NJ: John Wiley and Sons. (1971).

Argonne National Laboratory. *Natural Decay Series: Uranium, Radium, and Thorium* (Argonne National Laboratory Human Health Fact Sheet). (2005).Retrieved August 10, 2015 from: http://www-eng.lbl.gov/~shuman/NEXT/MATERIALS&COMPONENTS/Background_measurement/natural-decay-series.pdfhttp://www-eng.lbl.gov/~shuman/NEXT/MATERIALS&COMPONENTS/Background_measurement/natural-decay-series.pdf.

Bell, C.H., et al., Editors. *Emerging Contaminants Handbook.* Boca Raton, FL: CRC Press. (2019).

Biello, D. *Id Using Dispersants on the BP Gulf Oil Spill Fighting Pollution with Pollution? Scientific American.* (June 2010). Retrieved August 10, 2015 from: http://www.scientificamerican.com/article/is-using-dispersants-fighting-pollution-with-pollution/.

Biello, D. *Plastic (Not) Fantastic: Food Containers Leach a Potentially Harmful Chemical.* Scientific American. (2008). Retrieved August 7, 2015 from: http://www.scientificamerican.com/article/plastic-not-fantastic-with-bisphenol-a/.

Bienkowski B. *Exposure to metformin, a fist-line treatment for type-2 diabetes, feminizes male minnows and impacts fertility at levels common in wastewater effluent.* (2015). Retrieved August 8, 2015 from: http://www.environmentalhealthnews.org/ehs/news/2015/apr/diabetes-fish-endocrine-disruption-hormones-metformin.

Bini, C. and J. Bech, Editors. *PHE's, Environment, and Human Health.* Berlin: Springer. (2014).

Bower, J. *The Healthy House.* The Healthy House Institute. (1997).

Bower, M.L. *The Healthy Household.* The Healthy House Institute. (1997).

Boyles, W. *The Science of Chemical Oxygen Demand.* (Technical Information Series, Booklet No. 9). Hach Company. (1997). Retrieved August 8, 2015 from: www.hach.com/asset-get.download-en.jsa?code=61786.

Brown, T., LeMay, H., Bursten, B., et al. *Chemistry, the Central Science*, 12th Edition. Hoboken, NJ: Prentice Hall. (2012).

Captain Charles Moore with Cassandra Phillips. *The Plastic Ocean*; Avery/Penguin Group. (2011).

Chang, R. and Goldsby, K. *Chemistry*, 11th Edition. New York: McGraw Hill. (2013).

Chemical and Engineering News and Chemical Health and Safety; 1984–2015; Various Articles. Consumer Reports, Articles in November 2012 and January 2013 Issues.

Chemical and Engineering News *Special Edition*, "IT'S ELEMENTAL: The Periodic Table". (2003).

Clarke, R.H. and Valentin, J. *The History of ICRP and the Evolution of its Policies* (ICRP Publication 109) Elsevier. (2009). Retrieved August 10, 2015 from: http://www.icrp.org/docs/The%20History%20of%20ICRP%20and%20the%20Evolution%20of%20its%20Policies.pdf

Coderre, J. *Principled of Radioation Interactions.* [PDF document]. Retrieved August 10, 2015 from lecture notes online: http://ocw.mit.edu/courses/nuclear-engineering/22-55j-principles-of-radiation-interactions-fall-2004/lecture-notes/bakgrnd_radiaton.pdf.

DiCarlo, F. *Toxicology for Chemists-A Short Course.* Clearwater Beach, FL: American Chemical Society. (1986).

Editorial. Water wars. (2012). *Nature* 491, 496.

Elizabeth, G. *Chasing Molecules.* Washington, D.C.: Island Books|Shearwater Press. (2009).

Energy Information Administration. *International Energy Statistics.* (n.d.). Retrieved August 7, 2015 from: http://www.eia.gov/cfapps/ipdbproject/IEDIndex3.cfm?tid=1&pid=1&aid=2.

EPA Cleaner Power Plants. (2014). Retrieved August 7, 2015 from: https://19january2017snapshot.epa.gov/mats/cleaner-power-plants_.html

EPA Office of Air Quality. *Technical Bulletin: -Nitrogen Oxides (NOx), Why and How They Are Controlled.* (EPA 456/F-99-006R). (1999). Retrieved August 7, 2015 from: http://www.epa.gov/ttn/catc1/dir1/fnoxdoc.pdf.

EPA. A Citizen's Guide to Radon. (2013). Retrieved August 10, 2015 from: https://www.epa.gov/radon/citizens-guide-radon-guide-protecting-yourself-and-your-family-radon.

EPA. *Consumer's Guide to Radon Reduction.* (2013). Retrieved August 10, 2015 from: https://www.epa.gov/sites/default/files/2016-02/documents/2013_consumers_guide_to_radon_reduction.pdf

EPA. *Cross-State Air Pollution Rule (CSAPR).* (2015). Retrieved August 10, 2015 from: https://www.epa.gov/csapr/overview-cross-state-air-pollution-rule-csapr.

EPA. *Drinking Water Treatability Database Conventional Treatment.* (2015). Retrieved August 8, 2015 from: http://iaspub.epa.gov/tdb/pages/treatment/treatmentOverview.do?treatmentProcessId=1934681921.

EPA. *Overview of Greenhouse Gases: -Carbon Dioxide Emissions.* (2015). Retrieved August 8, 2015 from: http://www.epa.gov/climatechange/ghgemissions/gases/co2.html.

EPA. *Persistent Organic Pollutants: A Global Issue, A Global Response.* (2015). Retrieved August 8, 2015 from: http://www2.epa.gov/international-cooperation/persistent-organic-pollutants-global-issue-global-response.

EPA. *Reducing Acid Rain.* (2014). Retrieved August 10, 2015 from: http://www.epa.gov/air/caa/peg/acidrain.html.

EPA. *Water Health Series: Filtration Facts.* (2005). Retrieved August 10, 2015 from: https://www.epa.gov/sites/default/files/2015-11/documents/2005_11_17_faq_fs_healthseries_filtration.pdf

Favre, H.A. and Powell, W.H. *Nomenclature of Organic Chemistry. IUPAC. Recommendations and Preferred Name.* Cambridge, UK: The Royal Society of Chemistry. (2013).

Favre, H.A. and Powell, W.H. *Nomenclature of Organic Chemistry. IUPAC Recommendations and Preferred Name.* Cambridge, UK: The Royal Society of Chemistry. (2013).

Fowler, D., Flechard, C., Skiba, U., Coyle, M., Cape, J.N. The atmospheric budget of oxidized nitrogen and its role in ozone formation and deposition. (2008). *New Phytologist* 139, 11–23.

Gesser, H.D. *Applied Chemistry: A Textbook for Engineers and Technologists.* Dordrecht: Kluwer Academic/New York: Plenum Publishers. (2002).

Geyer, R., Jambeck, J.R. and Law, K.L. Production, use, and fate of all plastics ever made. (2017). *Sci. Adv.* doi:10.1126/sciadv.1700782.

"Ghosts of Bhopal" by Paul M. Barrett; Bloomberg Business, November 26, 2014.

*Acid Rain Program* Retrieved August 10, 2015 form Wikipedia: https://en.wikipedia.org/wiki/Acid_Rain_Program.

Gilbert, N. Drug-pollution law all washed up. (2012). *Nature* 491, 503–504.

Goldman, L. and Coussens, C. *Implications of Nanotechnology for Environmental Health Research*. Washington, DC: National Academic Press. (2005).

Gow, T., Pidwirny, M. *Land Use and Environmental Change in the Thompson-Okanagan*. (1996). Retrieved August 7, 2015 from: http://dwb4.unl.edu/Chem/CHEM869J/CHEM869JLinks/royal.okanagan.bc.ca/mpidwirn/atmosphereandclimate/smog.html.

Gushee, S.R. *The A-Z of D-Toxing: The Ultimate Guide to Reducing Our Toxic Exposures*. New York: The S File Publishing, LLC. (2015).

*Heavy Metal Pollution Is More Common Than You Think*. Retrieved August 8, 2015 from: http://www.fairfaxcounty.gov/nvswcd/newsletter/heavymetal.htm.

Hendricks, D. *Fundamentals of Water Treatment Unit Processes: Physical, Chemical, and Biological*. Boca Raton, FL: CRC Press. (2011).

Homburg, E. and E. Vaupel, Editors. *Hazardous Chemicals*. New York; Oxford: Berghahn Books. (2019).

Hugget, C., Levin, B. Toxicity of the pyrolysis and combustion products of poly (Vinyl Chlorides): A literature assessment. (1987). *Fire and Materials* 11, 131–142. Retrieved August 7th 2015 from: http://fire.nist.gov/bfrlpubs/fire87/PDF/f87015.pdf.

Hydrology Project. *Understanding chemical oxygen demand test* (Training Module # WQ-18). New Delhi May. (1999). Retrieved August 8, 2015 from: http://www.cwc.gov.in/main/HP/download/18%20Understanding%20COD%20test.pdf.

IPCC. *Climate Change 2013: The Physical Science Basis*. (2013). Retrieved August 7, 2015 from: http://www.ipcc.ch/report/ar5/wg1/.

Lide, D.R. *The Handbook of Organic Solvents*. Boca Raton, FL: CRC Press. (1995).

Lide, D.R., Editor. *Handbook Chemistry and Physics*, 101st Edition. London: CRC. (2020)

*Lime Softening* Retrieved August 10, 2015 from: http://water.me.vccs.edu/concepts/softeninglime.html.

McFadden, J. and Khalili, J. *Life on the Edge: The Coming of Age of Quantum Biology*. New York: Crown Publishers. (2014).

Mellow, J.W. *Intermediate Inorganic Chemistry*. London: Longmans, Green & Co. (1941).

Meyer, E. *Chemistry of Hazardous Materials*. Upper Saddle River, NJ: Pearson Education, Inc. (2014).

Myers, E. *Chemistry of Hazardous Materials*, 6th Edition. London: Pearson Educational. (2014).

Myers, S. and Frumpkin, H., Editors. *Planetary Health*. Washington & Covelo: Island Press. (2020).

Myers, S. and H. Frumpkin, Editor. *Planetary Health*. Washington, D.C.: Island Press. (2020).

Nalco *Oil Spill Dispersants*. (n.d.). Retrieved August 10, 2015 from: http://www.nalcoesllc.com/nes/corexit/oil-spill-dispersants.htm.

Olia, M., Editor. *Fundamentals of Engineering Exam*. 3rd Edition. Hauppauge, NY: Barron Educational Series. (2013).

Petrucci, R.H., Harwood, W.S. and Herring, G. *General Chemistry*, 8th Edition. Hoboken, NJ: Prentice Hall. (2002).

PMEL. *Ocean acidification: The other carbon dioxide problem* PMEL Carbon Program. (n.d.). Retrieved August 8, 2015 from: http://www.pmel.noaa.gov/co2/story/Ocean+Acidification.

Sánchez, C., Bøgesø, K.P., Ebert, B. et al. Escitalopram versus citalopram: The surprising role of the *R*-enantiomer. (2004). *Psychopharmacology* 174, 163–176

Smith, R. and B. Lourie. *Slow Death by Rubber Duck*. Berkeley, CA: Counterpoint| Berkeley. (2009).

Sobhani, Z., et al. Microplastics generated when opening plastic packaging. (2020). *Sci. Rep.* doi:10.1038/s41598-020-61146-4.

Spanu, A., et al., The Role of Irrigation Techniques in Arsenic Bioaccumulation in Rice (*Oryza sativa* L.). *Environ. Sci. Technol.* doi:10.1021/es300636d.

Stwertka, A. *A Guide to the Elements.* Oxford: Oxford University Press. (1998).

*The EU Emissions Trading System (Eu ETS)* Retrieved August 10, 2015 from: https://ec.europa.eu/clima/policies/ets_en.

Tian, Z., et al., A ubiquitous tire rubber–derived chemical induces acute mortality in coho salmon. (2020). *Science.* doi:10.1126/science.abd6951.

U.S. National Institutes of Health; *NIOSH Pocket Guide to Hazardous Substances.* (2007).

United States EPA.gov. Polychlorinated Biphenyls. John Wiley and Sons. Structure Elucidation in Organic Chemistry: The Search for the Right Tools. (2014).

US Department of Energy. *800,000-year Ice-Core Records of Atmospheric Carbon Dioxide* ($CO_2$). (2012). Retrieved August 7, 2015 from: http://cdiac.ornl.gov/trends/co2/ice_core_co2.html.

USGS. *Radioactive Elements in Coal and Fly Ash: -Abundance, Forms, and Environmental Significance* (Fact Sheet FS-163-97). (1997). Retrieved August 7, 2015 from: http://pubs.usgs.gov/fs/1997/fs163-97/FS-163-97.html.

USGS. *Water hardness and alkalinity.* (2013). Retrieved August 10, 2015 from: (13) http://water.usgs.gov/owq/hardness-alkalinity.html.

Valentin, J. *The 2007 Recommendations of the International Commission of Radiological Protection.* (2007). Elsevier Retrieved August 10, 2015 from: http://www.icrp.org/docs/ICRP_Publication_103-Annals_of_the_ICRP_37(2-4)-Free_extract.pdf.

Vallero, D. and Letcher, T. *Unraveling Environmental Disasters.* Amsterdam: Elsevier. (2013).

*Water Vapor and Saturation Pressure in Humid Air* (n.d.) Retrieved August 7, 2015 from: http://www.engineeringtoolbox.com/water-vapor-saturation-pressure-air-d_689.html.

*What is Nanotechnology?* Retrieved August 10, 2015 from: http://www.nano.gov/nanotech-101/what/definition.

*What is Water Pollution?* Retrieved August 8, 2015 from: http://www.conserve-energy-future.com/sources-and-causes-of-water-pollution.php.

WHO. *Handbook on Indoor Radon: A Public Health Perspective.* (2009). Retrieved August 10, 2015 from: http://apps.who.int/iris/bitstream/10665/44149/1/9789241547673_eng.pdf.

Woods Hole Oceanographic Institute. *Ocean acidification.* (n.d.). Retrieved August 8, 2015 from: http://www.whoi.edu/main/topic/ocean-acidification.

World Health Organization. *Dioxins and their effects on human health* (Fact sheet No 225). (2014). Retrieved August 8, 2015 from: http://www.who.int/mediacentre/factsheets/fs225/en/.

Zhang, F., et al., Polyethylene upcycling to long-chain alkylaromatics by tandem hydrogenolysis/aromatization. (2020). *Science.* doi:10.1126/science.abc5441.

# Index

Note: **Bold** page numbers refer to tables and *italic* page numbers refer to figures.

**Taylor & Francis Group**
an **informa** business

# Taylor & Francis eBooks

www.taylorfrancis.com

A single destination for eBooks from Taylor & Francis
with increased functionality and an improved user
experience to meet the needs of our customers.

90,000+ eBooks of award-winning academic content in
Humanities, Social Science, Science, Technology, Engineering,
and Medical written by a global network of editors and authors.

## TAYLOR & FRANCIS EBOOKS OFFERS:

A streamlined
experience for
our library
customers

A single point
of discovery
for all of our
eBook content

Improved
search and
discovery of
content at both
book and
chapter level

## REQUEST A FREE TRIAL
support@taylorfrancis.com

**Routledge**
Taylor & Francis Group

**CRC Press**
Taylor & Francis Group